Rydberg Series in Atoms and Molecules

RYDBERG SERIES IN
ATOMS AND MOLECULES

A. B. F. Duncan

Professor Emeritus of Chemistry
University of Rochester
Rochester, New York

Visiting Professor of Astronomy
University of Virginia
Charlottesville, Virginia

ACADEMIC PRESS   New York and London   1971

ACADEMIC PRESS, INC.
111 Fifth Avenue, New York, New York 10003

United Kingdom Edition published by
ACADEMIC PRESS, INC. (LONDON) LTD.
Berkeley Square House, London W1X 6BA

LIBRARY OF CONGRESS CATALOG CARD NUMBER: 75-159536

PRINTED IN THE UNITED STATES OF AMERICA

Contents

Preface

Rydberg states form an important class of excited electronic states of atomic and molecular systems. Although extensive reference to various aspects of these states are found in standard textbooks and research papers, there does not appear to exist any recent review or monograph devoted specifically to the subject. One purpose of the present book is to collect pertinent experimental data and to treat current theories of Rydberg states in one convenient place.

The older specific reference books on the subject treated atomic Rydberg states exclusively. Indeed, examples of molecular Rydberg series were few in number at the time of publication of these older monographs. However, the number of experimental examples has increased substantially since about 1934 and there has been a parallel but more moderate increase in understanding of the relevant theory. At the present time there is a renewed interest in high energy states of atoms and molecules which is stimulated by general problems of space research. It is hoped that this book may stimulate further progress in both experimental studies and in theoretical interpretation of results.

Selection of Rydberg states as a specific class of excited quantum states probably has no fundamental physical basis. Accordingly, a completely separate treatment of these states has no justification. No specific discussion can be made without acknowledgment to the excellent general treatments of excited states found in the publications listed in the general references. The publications of R. S. Mulliken and G. Herzberg were most helpful in the preparation of this volume, and I should like to acknowledge my indebtedness particularly to these sources.

The book is written at an elementary level in an attempt to meet the need of nonspecialists and students. Very little quantum mechanics and almost no mathematics are introduced, although exposure to an elementary course in the former discipline might be helpful. A thorough basic course in physical chemistry should be sufficient to understand all the material presented here.

Rydberg Series in Atoms and Molecules

1

Introduction

Regularity in distribution of lines in spectra of atoms was evident in the beginning of experimental spectroscopy. Attempts to find laws for these regularities have been described admirably in earlier reference works[1-3] and are of great historical interest. It will not be necessary in the present volume to record the early experimental results and interpretation, nor to give references to the original papers. It is appropriate to begin with some discussion of the spectrum of the hydrogen atom. At the same time, some elementary and useful concepts and definitions will be introduced.

In 1885, Balmer pointed out that a system of experimentally known lines in the spectrum of hydrogen could be represented by a formula which we write today as

$$1/\lambda = \nu = \text{const} \left[\tfrac{1}{2}^2 - 1/n^2\right]; \qquad n = 3, 4, \ldots . \qquad (1)$$

Here λ is the wavelength of light in centimeters and ν is the wave number, proportional to the frequency ν' (sec^{-1}), with $\nu' = c\nu$. The lines with $n = 3, 4$ are usually called H_α, H_β, etc., and lie in the visible region of the spectrum. As is obvious from the formula, the positions of the lines converge as n increases, and so form a spectroscopic series.

The formula of Balmer was generalized by Rydberg to include other observed wave numbers in the hydrogen spectrum to

$$\nu = \text{const} \left[(1/n_2{}^2) - (1/n_1{}^2)\right] \qquad (2)$$

with n_1, n_2, integers and $n_1 > n_2$. Spectroscopic series corresponding to $n_2 = 1, 3, 4, 5$ have been found, and each of these series also converges to $\nu = \text{const} (1/n_2{}^2)$ as n_1 approaches infinity. Within the accuracy of the experimental data, the constant, which we now call the Rydberg constant (R), has the same value for all the experimental series. This fact suggested to Rydberg and to Ritz that all observed wave numbers could be expressed as a difference in terms (T), each of the form R/n^2, so that

$$\nu = T_i - T_j .$$

Now the wave numbers emitted in a transition between two energy states is given by the general expression $hc\nu = E_2 - E_1$ so that $T_2 = E_2/hc$, $T_1 = E_1/hc$. However, the energy contains an additive constant which may be chosen by convention for tabulation of terms.

One may choose the additive constant so that at $E = 0$, the electron is separated from the system. The energy associated with the term then increases, in a negative sense, with decreasing n. Or one may choose the lowest energy state to correspond to $E = 0$ and measure energies increasing in a positive sense with increasing n. The latter choice is adopted in the most recent and complete tabulation of spectroscopic energy levels.[4] The former choice is, in many ways, more convenient, especially for comparison with theory. This choice will be adopted here. In either case, the term is taken to be positive and the energy of a bound state is clearly negative. It is also obvious that the difference in base of energy of the two choices represents the energy necessary to remove the electron—the ionization energy. Therefore spectroscopic terms are obtained from the tabulation of Moore[4] easily by subtraction from the appropriate series limit measured above the ground state of the atom.

The theoretical expression for the Rydberg constant, in terms of other fundamental constants of nature was deduced by Bohr[5] in 1913, and agreed with the experimental spectroscopic value within the probable accuracy of these fundamental constants. The Bohr theory of the hydrogen atom must be regarded as one of the most significant milestones in the progress of modern physics. The theoretical value of R is

$$R = 2\pi^2\mu e^4/ch^3,$$

where $\mu = Mm/(M + m)$; M is the mass of the nucleus and m is the mass of the electron; e, c, and h have their usual meaning. Therefore, R increases slightly with increasing nuclear mass to a limit

$$R(\text{infinite mass}) = 2\pi^2 me^4/ch^3.$$

The energies of the hydrogen atom are negatives of the terms R/n^2, and are

$$E = hc\nu = -(hc)R/n^2 = -2\pi^2\mu e^4/h^2 n^2.$$

In the ground state, $n = 1$ and

$$E_1 = -2\pi^2\mu e^4/h^2.$$

Also, when $n = \infty$, corresponding to removal of the electron, $E_\infty = 0$, and so $(-E_1)$ is the ionization potential of the hydrogen atom. According to the Bohr theory, $E = -e^2/2r$ and $r = n^2h^2/4\pi^2\mu e^2$. When $n = 1$, $r_1 = a_0 = h^2/4\pi^2\mu e^2$. The quantity a_0 is the atomic unit of distance. It is convenient also to adopt an atomic unit of energy. It follows that $E = -e^2/(2a_0 n^2) = -(R)hc/n^2$, and thus a convenient unit of energy is $Rhc = 2\pi^2\mu e^4/h^2$. This is the Rydberg atomic unit of energy. Using this unit, the energy states of the hydrogen atom are

$$E = -1/n^2.$$

It may be noted that the ionization potential of the hydrogen atom is exactly 1 Rydberg.

Another atomic unit of energy is the Hartree $2Rhc$. This unit is used more frequently in theoretical discussion. Using the Hartree, the hydrogenic energies are

$$E = -1/(2n^2).$$

The hydrogenic terms are $T(n) = 1/(2n^2)$ hartrees and thus are a function only of n, the principal quantum number. Energy levels characterized by different values of l and m, but the same n, are degenerate. If relativity effects are considered, there is a slight dependence of energy on l, in agreement with the observed fine structure of the spectral lines. The observed splitting is very small, and relativity effects will be neglected here and in further discussion.

The energy formulas are easily generalized to include other one-electron atomic systems (e.g., He^+, Li^{2+}, Be^{3+}), by insertion of a factor Z^2 in the numerators. Here, Z is the nuclear charge taken as 1 for H, 2 for He^+, 3 for Li^{2+}. With this modification, the formulas fit the observed Rydberg series in these species within the accuracy of observation. It is clear from the preceding discussion that all excited states of hydrogenic atoms are Rydberg states. In the case of a single valence electron outside a closed shell core, the excited states of the valence electron are usually classified as Rydberg states even when there is no increase in n on excitation. The classification is certain when the n is increased. When the core is not composed of closed shells then excitation without increase in n (as 2s–2p, in Be), is customarily described as a valence-shell transition, and not as a transition to a Rydberg state.

When more than one electron is present, the Schrödinger equation for the system cannot be solved exactly, and it is customary to seek one-electron solutions, or atomic orbitals, which satisfy a self-consistent field provided by the nuclear charges and other electrons. The state functions are then taken to be products of atomic orbitals. To satisfy the Pauli principle, the products are made properly asymmetrical by writing the state functions as determinants.

The nuclei and unexcited electrons constitute a core and the unexcited electrons are described by core atomic orbitals. An electron probability distribution may be obtained from the radial parts of the core orbitals in the usual way. Thus a core size can be defined as a volume containing an arbitrarily large fraction of the core electron density. The maximum in the radial part of the excited orbital will similarly define a radius and a size of this orbital. When the average radius of the excited orbital is much larger than the average radius of the core, the electron has been excited to a Rydberg orbital and a Rydberg state results.

In systems with more than one electron, energy levels with the same n but different l are not degenerate. One of the first examples studied was the spectrum of the sodium atom. Again, we do not discuss the historical aspects of these studies, but state the results, most of which are well known.

The electron structure of the sodium atom consists of a closed shell with a neon-like structure and a single (valence) electron. A similar model applies to the Li, K, Rb, and Cs atoms. The valence electron is certainly most easily excited and finally removed to give an ion with an approximate rare gas structure which acts as the core. However we expect distinct Rydberg series corresponding to excitation of the valence electron to ns ,np, nd, etc. levels, all converging to the lowest ionization potential. Such series are indeed observed with more or less completeness.

The series terms may be represented by a more general Rydberg formula

$$T = Z^2/2(n - \delta)^2. \tag{3}$$

Here, Z is taken as the *core* charge, and $Z = 1$ for neutral atoms. In this formula, the dependence of energy on l is taken into account through the quantity δ, the Rydberg defect. The Rydberg defect δ decreases strongly with increase in l, and becomes negligible in most cases with $l \geqq 3$. There is also a small dependence of δ on n. This dependence is included in more accurate formulas.

Representative numerical values of δ computed from experimental atomic terms are shown in Table I. An extended discussion of δ would be out of place in this introductory chapter, but it should be stated here that δ is related to ideas of penetrating orbits in the older quantum theory and elliptical forms of such orbits, depending on l. For a given l, δ appears to increase with increasing Z, since the Rydberg electron is penetrating into less tightly bound inner electron shells. For example, δ is smaller for Li, where the inner shell is $(1s)^2$, than for Na, for all corresponding values of l. In sodium, most of the penetration probably occurs in the $(2s, 2p)$ shell. Transitions are observed also for atoms with more than one-electron outside closed shells.

TABLE I

REPRESENTATIVE SPECTROSCOPIC TERMS AND δ_n IN SPECTRA OF
NEUTRAL ALKALI ATOMS

Atomic States	n	Term (cm^{-1})	δ
Li, ^{2}S	3	16281.07	0.404
	7	2519.29	0.400
	10	1189.19	0.394
Li, ^{2}P^0	2	28583.36	0.042
	7	2269.84	0.047
	10	1108.03	0.049
Li, ^{2}D	3	12204.09	0.0015
	7	2240.69	0.0021
Na, ^{2}S	4	15709.79	1.357
	7	3437.58	1.350
	10	1466.67	1.350
Na, ^{2}P^0	3	24493.47	0.883
	7	2909.25	0.858
	10	1312.42	0.856
Na, ^{2}D	3	12276.77	0.010
	7	2248.69	0.0143
	10	1100.48	0.0142

Here we must attempt to define a correction to the energy, through n, which includes penetration and exchange effects with both inner and outer parts of the core. These complications cause the concept of penetration to lose much of its value.

Experimental study of Rydberg series in molecules constitutes a relatively recent development in experimental spectroscopy. There are several reasons for this slow progress that will be noted here and discussed in more detail in later chapters. First, the ionization potentials of most simple molecules are relatively large, and series converging to them lie in the vacuum ultraviolet region of the spectrum. Second, the width of electronic levels is usually larger than in an atom because of vibrational and rotational motion. This situation leads to difficulty in resolution of the electronic levels, particularly when the transition appears as a broad continuous region. As the ionization limit is approached, transitions to Rydberg states become weaker, even though they may be sharp. Thus, they are difficult to find if there are stronger transitions present in the same region. Also the separation between electronic levels in a series becomes of the order of magnitude of separation of vibrational–rotational levels and ultimately the Born–Oppenheimer separability of electronic and nuclear motions is expected to become invalid, leading to difficulties in the theory of Rydberg transitions. A third reason arises from the fact that the electronic eigenfunctions depend on the nuclear configuration as well as electronic coordinates, and the electronic state energies have a parametric dependence on the nuclear separation R.

A most favorable starting point in consideration of molecular Rydberg series would seem to be the one-electron case of H_2^+. The quantum mechanical problem can be solved exactly and the electronic energy levels have been computed accurately over a wide range of R. No observation of experimental spectroscopic transitions has been reported and none is expected since all allowed transitions are to unstable upper states, leading to broad and weak continuous absorption at very short wavelengths. Accurate calculations have been made also for HeH^{2+}, but again no experimental spectra are available. Several fragmentary Rydberg series have been observed for H_2, but the number in any one series is insufficient to allow an accurate determination of the ionization potential from a series limit. Series in He_2 probably are the most extensive among the most simple diatomic molecules.

It is an empirical fact that observation of long Rydberg series in absorption is associated with small changes in the equilibrium internuclear distances of excited states R_e, relative to R_e of the normal state and the ion, particularly when the absolute value of R_e is small for all states. This situation corresponds to an almost vertical alignment of the potential curves for the excited states. It appears reasonable to assume simply from the Franck–Condon principle, that Rydberg transitions will appear with reasonable intensity in adsorption from low vibrational levels of the ground state. Furthermore, the Rydberg transitions should be accompanied by a minimum observable vibrational structure, with the result that the electric transitions should be well resolved and observed even at high values of n.

There are some difficulties in classification of excited states of these simple molecules into Rydberg and non-Rydberg states over wide ranges of R. The change of nature of such states with R has been discussed by Mulliken[6a-c] in a series of recent papers. Actually the number of diatomic molecules that exhibit well-established Rydberg series is small. Herzberg[7] lists only four or five, and the additions since 1950 are relatively small in number. This is not to say that individual Rydberg states of diatomic molecules occur infrequently, but only that the states associated with high values of n have not been reported. Therefore, the observed states cannot be fitted to a series formula.

Examples of Rydberg series in polyatomic molecules are much more numerous than in diatomic molecules. A number of reasons can be advanced in explanation of this situation: Perhaps the electronic structures of polyatomic molecules are of greater general interest in many areas of physical science—especially chemistry. Also the diatomic molecules whose stability suggests study in absorption have rather high ionization potentials, usually above 12 eV. This means that most of the higher members will be in spectral region where observations are difficult. A more fundamental reason would seem to be that lone pair or nonbonding orbitals are much more common in stable polyatomic molecules than in stable diatomic molecules. Long Rydberg series appear to result usually from excitation and final removal of such orbitals at the lowest ionization potential, with only slight changes in molecular bonding. However this reason is rather complex and will be discussed at length elsewhere.

The following chapters will present a more detailed account of Rydberg states. Consideration is given first to atomic examples in their experimental and theoretical aspects. Some principles that have been already pointed out briefly will be discussed further. The atomic case will be followed by chapters on diatomic and on polyatomic molecules, in which special characteristic new features of the theory will be introduced.

2

Rydberg Series in Atomic Spectra

2.1. GENERAL PRINCIPLES AND METHODS

Rydberg series in atoms are not only of interest in themselves but form models from which series in molecules can be understood. The theory of excited states of many electron systems is more highly developed in atoms than in molecules, principally because of the more simple nature of the atomic core. The essential complexities of molecular cores arise from motions of the nuclei and from differences in equilibrium internuclear distances in various excited states. Provided these equilibrium distances are small in comparison with the radius of the Rydberg orbit, many of the core properties can be deduced from

9

comparison to the corresponding united atom. The comparison is particularly useful in the polyatomic case of a central heavy nucleus with symmetrically bonded hydrogen nuclei, as in methane.

There are a number of theoretical methods for quantitative description of atomic structures. For one-electron systems, we may solve the Schrödinger equation exactly. Approximate methods are necessary for atoms with two or more electrons, but many of these methods give results of relatively high accuracy. Furthermore, in some methods, accuracy can be improved by higher-order approximations.

In all methods, we are seeking solutions of Schrödinger eigenvalue equations $H\psi = (T + V)\psi = E\psi$, where T is the kinetic energy of the system and the ψ are many-electron solutions. The approximate treatments differ essentially in the various methods of decomposition of V into a sum of potentials, and in the approximations made to the individual potentials in the sum. The approximate treatments are all concerned basically with the potential V of the many-electron systems, which we may write as

$$V(r_1, r_2, \ldots r_i, r_{i2}, \ldots r_{ij}).$$

For two-electron systems (He, Li^+, Be^{2+}), reliable results may be obtained by variational methods, perturbation methods, and combinations of the two. These methods and applications to He are described in detail by Bethe and Salpeter[8a] and in other general references.

The Ritz variational method is most useful for approximate computation of the lowest eigenvalue and corresponding wave function of the normal state of the atom. An analytical function ϕ is chosen, which contains a number of adjustable parameters, and is assumed to be an approximate solution to the exact Hamiltonian H of the system. Then ϕ satisfies the equation

$$H\phi = E\phi.$$

Then the *functional* $E(\phi)$ is given by the expression

$$E(\phi) = \int \phi^* H\phi \, dV$$

if ϕ is normalized. If ϕ were the exact normalized wave function of the normal state ψ_0 then $E(\phi) = E_0^0$ the exact eigenvalue. We now vary

$E(\phi)$ systematically with respect to the parameters of ϕ until a stationary value of $E(\phi)$ is obtained. The value will be an absolute minimum in this special case. Then it may be shown that $E(\phi) \geqq E_0^0$. The best solution ϕ consistent with the choice of parameters may bring $[E(\phi) - E_0^0]$ to a very small but always positive number, with a corresponding approach of ϕ to ψ_0, the exact wave function.

Since the treatment of V is fundamental and may be illustrated more clearly in perturbation methods, it appears best to discuss them firts.

The complete Hamiltonian for a system of i electrons (omitting spin terms and relativistic corrections) is (Hartree units)

$$H = \sum_i \left(-\frac{\nabla_i^2}{2} \right) + \sum_i (-Z/r_i) + \sum_{j>i} r_{ij}^{-1}.$$

The second and third terms constitute V; Z is the nuclear charge. In a perturbation method, we split H into $H = H_0 + \lambda H_1$, the perturbation to H_0. The objective in all perturbation methods is choose V so that λH_1 is small and at the same time to allow a simple solution of a Schrödinger equation

$$(H_0 + \lambda H_1)\psi = E\psi. \tag{2-1}$$

Regardless of the details of partition of V between H_0 and H_1, higher orders of perturbation will lead, in principle, to improved values of ψ and E, although the process may become too difficult to carry through in practice.

The expansions

$$\psi = \sum_{n=0}^{\infty} \lambda^{(n)}\psi_n, \qquad E = \sum_{n=0}^{\infty} \lambda^{(n)}E_n,$$

where n is the order of perturbation, are inserted into Eq. (2-1) and coefficients of each pair of λ are equated. This leads to coupled differential equations connecting E_n, ψ_n of various orders n of perturbation.

(a): $\qquad\qquad H_0\psi_0 = E_0\psi_0,$

(b): $\qquad\qquad H_0\psi_1 + H_1\psi_0 = E_0\psi_1 + E_1\psi_0,$

(c): $\qquad\qquad H_0\psi_2 + H_1\psi_1 = E_0\psi_2 + E_1\psi_1 + E_2\psi_0.$

Contrary to usual notation, the subscripts in these equations refer to the order of perturbation; each equation has, of course, a family of particular orthonormal solutions and corresponding eigenvalues.

The solution of the coupled equations proceeds as follows. We regard E_0 and ψ_0 as known, multiply (a) by ψ_1 and (b) by ψ_0, subtract, integrate the result, and obtain

$$E_1 = \langle \psi_0 H_1 \psi_0 \rangle.$$

The first-order perturbation energy associated with a particular ψ is thus obtained from ψ_0, the corresponding unperturbed function. Since E_1 has been found, Eq. (b) can be solved for ψ_1. In a similar way, multiply (a) by ψ_2 and (c) by ψ_0, subtract, integrate, and obtain the second-order perturbation energy

$$E_2 = \langle \psi_0 H_1 \psi_1 \rangle,$$

and ψ_2 can now be obtained from (c).

This formal procedure can be extended to obtain eigenvalues and eigenfunctions in higher orders of perturbation. While this general method seems conceptually very simple, the actual process becomes very complicated for ψ in higher order than one and so for E in higher order than three. This is because ψ_1 must be obtained for the solutions of the above differential equations. In the first place, ψ_0 for any particular ψ must be known rather accurately otherwise large errors may arise in E_1 and become larger in ψ_1, E_2, and ψ_2. Secondly, any particular ψ_1 must be regarded as an expansion into a complete set of ψ_0, and therefore *all* of the matrix elements $\langle \psi_0 H_1 \psi_0 \rangle$ of H_1 must be calculated if E and ψ_1 are to be determined. The complete set of zero-order functions must include *all* solutions of H_0, including the ψ_0 which belong to continuous values of the zero-order energies. The contribution from the continuous values may be very large, and is difficult to assess. Difficulties increase as one proceeds toward E_2 and ψ_2.

The formal difficulties in determination of ψ_1 may be avoided partly for a particular ψ, by introduction of variation theory—a method suggested by Bethe and Salpeter.[8b] Suppose we know E_0, ψ_0, and E_1, and we are content to obtain an approximate ψ_1 for the same state. A trial function ϕ_1 is assumed with arbitrary parameters so that $E_2(\phi_1)$ may be varied with respect to the parameters. A minimum value of $E_2(\phi_1)$ will be an upper limit for E_2, while $E_2(\phi_1)$ will be a good approximation to E_2, provided ϕ_1 is sufficiently flexible, and so ϕ_1 may approach ψ_1 closely. This variational–perturbation method may be extended with the approximate ψ_1 to obtain E_3, ψ_2, and E_4, although errors may be expected to accumulate in ψ_2, because of inaccuracies in the approximate ψ_1.

Variational calculation of functions for excited states usually is made subject to conditions of restraint. The trial function is required to be orthogonal to the exact wave functions of all lower states. This requirement is satisfied automatically for the lowest excited states, which have orbital symmetry, multiplicity, or both different from the normal state. The requirement is not satisfied automatically for the lowest excited 2^1S state of He, for example.

A variational treatment of the 2^1S state may be carried out, however, by the linear variation method. The trial function is expressed as a linear sum of terms of the correct symmetry containing (nonlinear) parameters in exponential factors. Variation only with respect to the coefficients of the terms leads to a secular equation. The lowest root corresponds to an approximate ground-state energy. The next highest root will be an approximation to the energy of 2^1S, which must always lie higher than the experimental energy. Then variation with respect to the nonlinear parameters may be made until a stationary value of the second root, or energy of 2^1S is obtained. This procedure may be applied to higher states such as 3^1S, 4^1S, and the stationary values of successively higher roots are each above the corresponding exact, or experimental energies.[9]

A more simple treatment, but of somewhat questionable validity assumes that the trial function $\phi_1(2^1S)$ has some symmetric orbital form such as $[1s(1)2s(2) + 2s(1)1s(2)]$ multiplied by an antisymmetric spin function. Likewise there exists a trial function $\phi_0(1^1S)$ for the same form $[1s'(1)1s'(2)]$. Provided that $1s' = 1s$, then $\langle\phi_0\phi_1\rangle = k\langle 1s2s\rangle$. Thus if ϕ_1 and ϕ_0 are required to be orthogonal, then $\langle 1s2s\rangle = 0$, and the reverse is also true. If $1s' \neq 1s$ in ϕ_0, perhaps because of different effective nuclear charge, then ϕ_0 and ϕ_1 will be orthogonal only if it is required that $\langle 1s'2s\rangle = 0$. Then, in general, $\langle 1s2s\rangle \neq 0$.

Orthonormality of trial functions is very important in the general consideration of excited states. If no orthogonality requirement is imposed, one usually will not obtain a stationary value of the energy of an excited state. A stationary value usually is obtained when conditions of orthogonality are imposed but there is no guarantee that the stationary energy will be higher than the exact energy of an excited state and so represent an upper bound to the energy. The stationary value may, in fact lie below the exact energy and thus the trial function forms a bad approximation to an exact eigenfunction of a lower state.

This point may be discussed further in the example of $E(2^1S)$ of helium. Let $E(2^1S) = E_1(\phi_1)$, $E^0(2^1S) = E_1^0(\psi_1)$ and $E^0(1^1S) = E_0^0(\psi_0)$. Then the variational theorem will be satisfied if $E_1 - E_1^0 \geqq 0$. The conditions to be met may be found in a formal way[10] through expansion of the trial functions, ϕ_1 for the (2^1S) into a complete set of exact eigenfunctions ψ_n. Since the ψ_n are necessarily orthonormal and $\langle \psi_n H \psi_n \rangle = E_n^0$,

$$\phi_1 = \sum_{n=0}^{\infty} a_{1n} \psi_n, \qquad (2\text{--}2)$$

$$E_1 = \langle \phi_1 H \phi_1 \rangle = \sum_{n=0}^{\infty} a^2_{1n} E_n^0. \qquad (2\text{--}3)$$

Subtracting E_1^0 from both sides of Eq. (2–3), and making use of the fact that ϕ_1 is normalized, so that $\sum_{n=0}^{\infty} a^2_{1n} = 1$, we have, in general,

$$E_1 - E_1^0 = \sum_{n=0}^{\infty} a^2_{1n} (E_n^0 - E_1^0) \geqq 0.$$

Therefore, the variational theorem will be satisfied for E_1 only *if* $\sum_{n=0}^{\infty} a^2_{1n} (E_n^0 - E_1^0)$ is positive.

In the particular case under consideration, the summation may be rewritten

$$a^2_{10}(E_0^0 - E_1^0) + \sum_{n=2}^{\infty} a^2_{1n} (E_n^0 - E_1^0) > 0.$$

The inequality is satisfied, or course, if $a_{10} = 0$, since the summation is necessarily positive. However, if $a_{10} \neq 0$, we have

$$a^2_{10}(E_1^0 - E_0^0) < \sum_{n=2}^{\infty} a^2_{1n} (E_n^0 - E_1^0),$$

where now both sides of the inequality are positive. Although $(E_1{}^0 - E_0{}^0)$ itself is large, it may be possible in specific cases that a_{10} is sufficiently small to meet this inequality, and if so, the variational process is justified. It is further apparent from the expansion of ϕ_1 that $a_{10} = \langle\phi_1\psi_0\rangle$, so that the magnitude of a_{10} may be investigated in a qualitative way. Suppose we have obtained a good trial function ϕ_0 for the normal state, which is orthogonal to ϕ_1. Although we do not know ψ_0, it seems reasonable to assume that $\langle\phi_0\psi_0\rangle$, the overlap integral between trial and exact functions, approaches unity, as ϕ_0 becomes more nearly equal to ψ_0. If this is so, the coefficients a_{0n} in the expansion of

$$\phi_0 = a_{00}\psi_0 + \sum_{n=1}^{\infty} a_{0n}\psi_n$$

will be very small in comparison with a_{00}, which will be about equal to unity for a good ϕ_0. Consequently,

$$\langle\phi_1\phi_0\rangle \simeq \langle\phi_1\psi_0\rangle = a_{10},$$

and since ϕ_1 is or has been made orthogonal to ϕ_0, $a_{10} \simeq 0$.

A method for estimating the closeness of ϕ_0 to ψ_0 has been proposed by Eckart.[11] A measure of the error ϵ in ϕ_0 is given by the expression

$$\epsilon = \int |\phi_0 - \psi_0|^2 \, dV$$

By an expansion of ϕ_0 into a series of exact eigenfunctions, similar to Eq. (2-2),

$$\langle |\phi_0 - \psi_0|^2\rangle = 2(1 - a_{00}) \leqq (E_0 - E_0{}^0)/(E_1{}^0 - E_0{}^0),$$

where $E_0 = \langle\phi_0 H\phi_0\rangle$, $E_0{}^0 = \langle\psi_0 H\psi_0\rangle$, and $E_1{}^0 = \langle\psi_1 H\psi_1\rangle$. The quantity $1 - a_{00}$ is thus a measure of contributions of wave functions of higher states to the trial wave function.

This variation method may be applied in principle to higher states of the same symmetry, although the errors are expected to increase because of the approximations. It is to be noted that this method would fail completely without requirement of orthogonality of all ϕ_n to *all* trial functions of lower states with the same symmetry.

Shull and Löwdin[12] have extended this method to higher states, and have provided a variation theorem for excited states. Orthonormal approximate real functions ϕ_0, ϕ_1 are selected to satisfy the conditions

$$\langle \phi_0 \phi_1 \rangle = 0 \quad \text{and} \quad \langle \phi_0 H \phi_1 \rangle = 0.$$

Then it is easily seen that the mean square error of the excited state

$$\langle | \phi_1 - \psi_1 |^2 \rangle = 2(1 - a_{11}).$$

From the expansion of ϕ_0 and ϕ_1 into exact wave functions, it follows that

$$2(1 - a_{11}) \leqq [(E_1 - E_1^0) + a^2_{10}(E_2^0 - E_0^0)]/(E_2^0 - E_1^0).$$

Shull and Löwdin show further that under the condition that $E_1^0 > E_0$,

$$2(1 - a_{11}) \leqq [(E_1 - E_1^0)/(E_2^0 - E_0^0)]$$
$$\times [1 + (E_2^0 - E_0^0)/(E_1^0 - E_0)],$$

or

$$(E_1 - E_1^0) \geqq 2(1 - a_{11})(E_2^0 - E_1^0)$$
$$\times (E_1^0 - E_0)/[(E_2^0 - E_0^0) + (E_1^0 - E_0)].$$

Since $0 \leq a_{11} \leq 1$, and $(E_1^0 - E_0)$ is assumed to be positive, the right-hand side is positive. Thus,

$$(E_1 - E_1^0) \geqq 0,$$

and an upper bound for $E_1(\phi_1)$ is established. To prove that E_1 is actually an upper bound to E_1^0, the validity of the condition $(E_1^0 - E_0)$ must be established for approximate functions. These functions must also meet the initial expansion. A somewhat different treatment of excited states has been proposed by Sinanoğlu[13] based on variational and perturbation methods. Ordinarily, a perturbation calculation of the first-order wave function requires knowledge of all the zero-order (unperturbed) wave functions, including those associated with continuous values of the energy. Sinanoğlu shows how the first-order wave function for a particular state may be determined variationally, provided that the trial function is orthogonal to zero-order functions of all states of lower energy, either by symmetry or multiplicity. If the trial function is not orthogonal, an arbitrary part of it must be maintained orthogonal to all the zero-order functions for lower states. The trial function must then be chosen to conform to this restriction. For further details, the original publications should be consulted.

The form of the variation method just described becomes awkward for the simultaneous treatment of a number of excited states of many-electron systems. For this application, the self-consistent field (SCF) method, originally developed by Fock[14] and Hartree,[15] is more frequently used. The method is described in detail in the original papers, and we shall only outline the procedure.

The wave function is constructed as a product of one-electron (orbital) functions. In the original Hartree calculations, a simple product was used and exchange repulsion of the electrons was neglected. An antisymmetrized product was used by Fock.[14] An initial set of orbitals ϕ_i is chosen, one for each electron. The potential energy of one electron is calculated in the field of the nucleus and in the averaged field of all other electrons. The calculation is repeated for each electron. One-electron Schrödinger equations are then solved for each orbital in the field just calculated, and a new set of orbitals obtained, which, in general, will not be identical to the initially assumed set. The potential energy is then recalculated and the procedure continued iteratively until a set of orbitals agrees with the set from which the potential field was calculated in the previous state of iteration. The final result is a set of orbitals that satisfies a potential field which is calculated from the same orbitals. These orbitals are the best obtainable with an assumed form.

The atomic orbital functions ϕ, of course, are not strictly of a central field type because of angular as well as radial dependence. In atomic calculations, however, each ϕ can be written as a product of radial and angular factors. Integrations over the angles are carried out separately and so it is necessary to solve equations that involve only radial functions. The Hartree–Fock equations may be derived from the variation method as is shown by Hartree[15] and by Slater.[16]

Originally the Hartree–Fock equations were solved by numerical integration and the orbital functions are obtained as numerical values of $\phi(r)$, tabulated against r. Methods have been devised to fit such orbital functions to analytical forms. The Hartree–Fock method may be applied to any atom in any state. The original Hartree–Fock method allows calculation of zero-order eigenfunctions and first-order energy with very good accuracy, but it is not possible to obtain higher approximations by perturbation theory.

More recently, other procedures, known as expansion methods have been developed by Roothaan[17] and others for the solution of the Hartree–Fock equations. These methods were developed originally for the treatment of molecular systems, where numerical integrations would be impractical. The expansion methods may be applied to atomic systems equally well, and have the advantage of yielding the best orbitals in analytical forms, which may be used in calculations on molecules. The method is adapted to large-scale electronic computing, and the basic principles have been described by Roothaan and Bagus,[18] who give details of programs for machine calculation as well as specific examples of some results. We can give here only a very brief summary of the principles.

The total wave function is taken to have the correct symmetry and spin properties and is therefore expressed as a determinant or sum of determinants of spin–orbitals. In the expansion method, each orbital ϕ is expanded as a sum of basis functions, and each basis function is a product of radial and angular functions. The initial ϕ are orthonormalized. The energy of the determinantal function is then computed by application of variational methods and the "best" orbitals obtained by alternative procedures, as was described earlier. The procedure does not differ in fundamental principle from the original method of Hartree and Fock, but has added flexibility in the use of a larger number of radial functions and in a wider choice in the detailed forms of the radial functions. The basis set of radial functions R_{nl} usually are chosen to have exponential forms of the Slater type, with screening parameters (orbital exponents), ζ_{nl}. Energy integrals over such functions may be obtained easily from expressions in closed form. The total energy is obtained as a sum of one- and two-electron integrals and is varied with respect to the expansion coefficients, subject to the condition that the ϕ remain orthonormal, until a stationary value is reached. The iterative process may converge only slowly, and occasionally is found to diverge. Extrapolation methods, which increase the rate of convergence and which usually correct cases of divergence, are described by Roothaan and by Hartree.

The quality of the results increase with the size of the basis set, but the size or number of radial functions should be kept small to avoid complexities in computing programs and to reduce cost of computing. If the size is to be small, then the radial functions should be optimized. This is done by variation of the energy with respect to the nonlinear parameter, or orbital exponents ζ_{nl} in the basis functions. This variation is tedious but is usually more practical than use of very large basis sets. Optimization will change the coefficients of the expansion because the optimized functions must remain normalized and mutually orthogonal. It is then necessary to repeat the SCF procedure with each chosen set of nonlinear parameters.

We now return to further discussion of elementary perturbation theory, particularly to treatment of excited states. Difficulties in proceeding from first order to higher orders of perturbation have already been mentioned as arising from determination of first-order wave functions. The formal method consists in solving the zero-order equation

$$H_0\psi_0 = E_0\psi_0$$

for all possible zero-order eigenvalues and wave functions. Then the first order energies E_1 are the integrals

$$E_1 = \langle\psi_0 H_1\psi_0\rangle.$$

The values E_0, ψ_0, and E_1 can thus be considered as known, and we must next find all of the first-order functions ψ_1, in order to obtain the second-order perturbation energy, E_2.

Some of the early examples of two-electron problems quoted by Bethe[8a] are instructive. We mention first Heisenberg's treatment of excited states of helium. It is supposed that the unexcited electron, on the average close to the nucleus, is not screened by the outer excited electron and thus experiences a potential $V_1(r_1) = -Z/r_1$. On the other hand, the outer electron is completely screened from the nucleus by the inner electron and thus experiences a nuclear change of $(Z - 1)$, and $V_2(r_2) = -(Z - 1)/r_2$. The perturbing potential is $\lambda H_1 = (1/r_{12}) - (1/r_2)$. The zero-order wave functions are hydrogenic with different effective nuclear changes for the two electrons. The zero-order energy is

$$E_0 \simeq -Z^2/2 - (Z - 1)^2/2n^2,$$

in atomic units. The energy does not depend on l, and furthermore, is

independent of the multiplicity (singlet or triplet) of the states. It is
necessary to go to first-order perturbation theory to obtain these de-
pendencies.

The Hamiltonian may be split in two ways:

$$H = H_{0a} + \lambda H_{1a} = H_{0b} + \lambda H_{1b}.$$

The zero-order and perturbation parts are divided asymmetrically as

$$H_{0a} = -(\nabla_1^2/2) - (\nabla_2^2/2) - Zr_1 - (Z-1)/r_2^2,$$

$$\lambda H_{1a} = (1/r_{12}) - (1/r_2),$$

with a similar division of H_{0b}, with r_1, r_2 reversed in the formulas.
We consider zero-order eigenvalues E_{0a} of H_{0a} and E_{0b} of H_{0b} with
$E_{0a} = E_{0b} = E_0$. Bethe obtained the first-order energy, E_1 in this
formulation as

$$E_1 = \tfrac{1}{2} \int (\psi_{0a} + \psi_{0b})(H_{1a}\psi_{0a} + H_{1b}\psi_{0b}) \, dV.$$

The zero-order wave functions will have the form

$$\psi_0 = [\psi_{0a}(1)\psi_{0b}(2) \pm \psi_{0b}(1)\psi_{0a}(2)]/\sqrt{2}, \qquad (2\text{--}4)$$

where (1) and (2) refer to two electrons, plus corresponds to a singlet
state, and minus to a triplet state. It is to be noted that the ψ_0 are degenerate in zero order. The corresponding energies of singlet and triplet
are therefore equal. Furthermore, the zero-order energy does not depend on l. A splitting into singlet and triplet states, and a dependence
of E on l will occur in first order because of the presence of r_{12}^{-1} in the
perturbation.

The first-order energy is evaluated by taking analytic forms for the
zero-order function with a nuclear charge Z for ϕ_1, the inner electron
(1), and $Z - 1$ for the outer electron ϕ_{nlm}. Hydrogenic forms are
suitable. In the integrations, it is assumed that the zero-order functions are orthogonal in zero-order. The final result is

$$E_1 = \bar{J} \pm K,$$

where

$$\bar{J} = \langle \phi_1^2(1) \mid r_{12}^{-1} - r_2^{-1} \mid \phi^2_{nlm}(2) \rangle.$$

The bar on $\bar{J}$ series is to distinguish it from a similar integral

$$J = \langle \phi_1^2(1) \mid r_{12}^{-1} \mid \phi^2_{nlm}(2) \rangle,$$

usually defined as a Coulomb integral. The familiar exchange integral K is defined as

$$K = \langle \phi_1(1)\phi_{nlm}(1) \mid r_{12}^{-1} \mid \phi_1(2)\phi_{nlm}(2) \rangle.$$

In the helium atom,

$$\phi_1 = R_{10}(r)Y_{00}(\theta, \phi), \qquad \phi_{nlm} = R_{nl}(r)Y_{lm}(\theta, \phi).$$

The total energy of the two-electron systems, to the first order is thus

$$E_0 + E_1 = -Z^2/2 - (Z-1)^2/2n^2 + \bar{J} \pm K.$$

General formulas for $\bar{J}$ and K would be extremely complicated. Formulas for $n = l + 1$ were given by Heisenberg and are quoted by Bethe.[8] Asymptotic formulas for very large n, which show that $\bar{J}$ and K are proportional to n^{-3}, may be derived.

The total energy of the outer electron, to first-order perturbation is

$$-(Z-1)^2/2n^2 + \bar{J} \pm K.$$

The total energy is of the form of the Rydberg energy expression

$$E(n) = (Z-1)^2/2(n-\delta)^2 = -(Z-1)^2/2n^2 - (Z-1)^2\delta n^{-3}.$$

Since $\bar{J}$ and K are proportional to n^{-3}, δ is determined by $\bar{J} \pm K$ in first-order perturbation theory.

Furthermore, the Rydberg defect δ may be separated into contributions from $\bar{J}$ to K, which we label δ_J and δ_K. If the observed defects for singlet and triplet terms are $^1\delta$ and $^3\delta$, then $\delta_J = (^1\delta + {}^3\delta)/2$, $\delta_K = (^1\delta - {}^3\delta)/2$. Bethe and Salpeter[19] shows a tabulation of observed and calculated values of δ_j, δ_K, for several terms of He and Li$^+$. Agreement for the exchange part δ_K is good, except for S-terms. The computed Coulomb part δ_J is in poor agreement with experiment; it is necessary to go to second-order perturbation theory to improve the agreement. There are several conclusions that may be drawn from a study of the tabulated values and from the detailed equations given by Bethe. First, both δ_J and δ_K are almost independent of n, but decrease strongly with increase in l. The decrease of δ_J with increase of l may be ascribed to decreased penetration of the outer electron into the charge distribution of the inner (core) electron. The probability that an electron is at a distance r from the nucleus is proportional to $(r)^{2l}$. A similar decrease of δ_K with l occurs because smaller overlap of the distributions of the outer and inner electrons, with increasing l of the outer electron.

First-order perturbation theory requires only zero-order wave functions. These zero-order functions have been chosen as antisymmetrized products of independent single-electron functions. Any effect of the wave function of the outer electron (2) on the inner electron (1) has been ignored. This neglect is unrealistic on physical considerations, as has been pointed out by Bethe. The average kinetic energy of electron (2) is smaller than for the inner electron and therefore at each fixed position r_2, the rapidly moving inner electron experiences the zero-order potentials plus an additional potential due to the presence of the outer electron at a fixed distance. We may say that the outer electron polarizes the inner electron. The magnitude of the effect of polarization may be obtained from consideration of an extra potential $\epsilon(r_2)$, integrated over the coordinates of electron (1). The total energy will thus contain a part due to this polarization arising for the potential $\epsilon_2(r_2)$. An approximation to the second-order energy of the outer electron will be

$$E_2(2) \;=\; \int \phi_{n\,lm}(2) \mid \epsilon_2(r_2) \mid \phi_{n\,lm}(2) \; dV_2 \,.$$

Calculation of the potential $\epsilon_2(r_2)$ is described by Bethe and Salpeter[3a]. He has obtained the result for a simple product zero-order function and thus the polarization correction to $\bar{J}$ only has been obtained. For large values of n, $E_2(2)$ is proportional to n^{-3}. This is sufficient for calculation of a Coulomb "polarization" correction δ_π to δ_J.

The correction for polarization turns out to be rather large and the sum $\delta_\pi + \delta_J$ is in fairly good agreement with the experimental $\delta_J = (^1\delta + {}^3\delta)/2$, for all l values. There is also a very small polarization correction to δ_K, which would appear if an antisymmetrized zero-order function were used. However, the correction to δ_K for polarization is too small to remove the large disagreement with experimental δ_K for S states. Further analysis of this error is necessary.

The larger error in δ_K for S states is connected with larger penetration of outer (ns) electrons into the core, or into charge distribution of the inner electron. In Heisenberg's method, however, the outer electron is fully screened from the nuclear charge by the inner electron, with an effective nuclear charge $Z - 1$. Improvement may be obtained by Fock's method,[14] which attempts to find by a variational method the best forms of one-electron functions ϕ_1, ϕ_2 in a wave function

$$\psi = [\phi_1(1)\phi_2(2) \pm \phi_2(1)\phi_1(2)]/\sqrt{2}.$$

Variation of the orbitals ϕ_1, ϕ_2 leads to coupled linear equations that are solved interatively by numerical methods. Calculations were carried through by Smith[20] for S-states with large n. These calculations gave δ values for both singlet and triplet states that were in close agreement with experiment.

The most accurate wave functions for excited states have resulted from variational treatments, based fundamentally on the Ritz method. Whereas most of the calculations have been on two-electron systems, it is of interest to see how far they can be carried through, in the hope of extension to large systems, and to assess the importance of approximations. The calculations of Hylleraas[21] on the ground state of He, carried out in the early days of quantum mechanics, gave values of the energy that are only about 0.0005 Hartree units higher than the experimental value. Calculations on some of the lower excited states gave results that are only slightly inferior. Most of these calculations were made more than 30 years ago, and subsequent refinements have given only very slightly better values of the energy.

We wish to discuss, at least in outline, the method by which these accurate results were obtained. Some facts should be pointed out, although they must be obvious to most readers. All excited states of helium are based on open shell configurations, such as $(1s)(ns)$, $^{1,3}S$; $(1s)(np)$, $^{1,3}P$. Precise calculation of $(1s)(ns)$, ^1S is most difficult in a variational treatment because of restricting conditions of orthogonality, as has been mentioned before. We have not mentioned directly the importance of electron correlation, which may be exhibited in several ways. First there is no natural correlation in the motion of two electrons that have different spin functions, as in singlet states. Second, there is no natural angular correlation for electrons with the same l-value. There may be some radial correlation because of nodal surfaces in the radial wave functions for electrons with the same l but different n. Although these correlation types have been formally distinguished, the general effect of correlation is to keep electrons apart in space.

An obvious method of introducing correlation into a wave function is to include functions of the interelectronic distance r_{ij} as coordinates of the total wave function, so that the wave function is small when r_{ij} is small. Most of the best wave functions include the coordinate r_{ij}; usually the first power of r_{ij} is sufficient. A second but less obvious way is to base the wave function on a number of configurations. If the set of basis functions were infinite, then the wave function expanded into this infinite set would be exact. We must be satisfied with a truncated set in practice and unfortunately there is no *a priori* way to choose a small finite set that will converge rapidly. The best that can be done is to use the formal classification of correlative effects as a physical guide, and to examine the final result as the number of terms of the set is increased. The best variational wave functions have been obtained by use of large basis sets along with some function of r_{ij}.

When a trial function of the above-mentioned nature is used in a linear variation method, a secular determinant results. The eigenvalues or the roots of the determinant are ordered so that each root is guaranteed to be an upper limit to the corresponding exact eigenvalue.[9]

The trial radial function for the two-electron system is most conveniently expressed in coordinates, introduced by Hylleraas as

$$s = r_1 + r_2 ; \qquad t = r_1 - r_2 ; \qquad u = r_{12} .$$

The Hamiltonian is an even function of t and so only even powers of t are allowed in the expansion of the wave function. The trial function has the general form

$$\phi = \sum_{nlm} c_{n,\,2l,\,m} s^n t^{2l} u^m \exp(-\alpha s).$$

The exponential factor assures the correct behavior of ϕ at infinity. In this factor $\alpha = (Z - \sigma)$ and represents a partially screened nuclear charge; σ is a screening parameter.

The expectation value of the energy is

$$E = \langle \phi H \phi \rangle / \langle \phi \phi \rangle,$$

which is to be minimized with respect to the $C_{n,\,2l,\,m}$ and α.

The series terms are computed for a number of fixed values of α.

The matrix elements of H are constructed and the secular equation solved for the eigenvalues. The process is repeated for a series of fixed values of α and the minimum is determined by interpolation. The process is extremely tedious for a large number of expansion terms. As many as 38 terms have been used.

If only the first (000) term of the series were taken, the well-known result $\alpha = \frac{5}{16}$ is obtained. This case corresponds to an approximation of the wave function as a product of two hydrogenic functions, each screened from the nuclear charge Z by $\alpha = \frac{5}{16}$. The energy of the normal state in this most simple approximation is -2.848 a.u. (experimental value is -2.903746).[22] Three- and six-parameter functions of Hylleraas gave values of -2.90244, -2.90324, respectively. The 38-term function of Koshinita gave -2.903722. When small relativistic and other corrections are applied the energy is within 1×10^{-6} Hartrees or about 0.2 cm^{-1} of the experimental value.

A function such as (2–5) does not actually show us from what configurations of electrons it is built. One must analyze the function further in order to obtain any information on what configurations are most important in lowering the energy. If one wishes further information on details of electron correlation, it is necessary to construct a many-term function in which the terms represent contributions for definite configurations. Taylor and Parr[23] have studied this problem. Beginning with the configuration $(1s)^2$ they have added systematically configurations $(1s, 2s)(1s, 3s)(2s)^2$, and configurations containing p, and d electrons. After each addition, the total energy is computed. No factors dependent on r_{12} were included in their wave functions. The lowest energy reached is -2.89743 a.u., with configurations $(1s1s')$, $(2p)^2$, $(3d)^2$, $(4f)^2$. Their configuration $(1s1s')$ represents two 1s functions with different screening constants and therefore indicates a radial separation and radial correlation of the inner electrons. When $(1s1s')$ in this set of configurations is replaced by $(1s)^2$, they obtained a value of the energy 0.02555 a.u. higher. Furthermore the difference in total energy between the configurations $(1s1s')$ and $(1s)^2$ alone is 0.02800 a.u. The effect in the radial correlation in splitting the inner shell configuration is obviously large.

We discuss next some of the more accurate results on excited states of He obtained by the variation method. Calculation of $2\,^3S$, $2\,^1P$, $2\,^3P$, $3\,^1D$, $3\,^3D$ can all be made by the Ritz method without requiring any subsidiary conditions of orthogonality to lower states. Trial functions of the general type [Eq. (2–5)] of the proper symmetry have been used. For the 2 ^{3}S state Hylleraas and Underheim[24] used an antisymmetric function

$$\phi = e^{-\alpha s}(c_1 + c_2 s + c_4 u + c_5 su) \sinh \beta t + t(c_3 + c_6 u) \cosh \beta t$$

where s, t, and u have been defined before and α and β are sum and difference of nuclear charges for the two electrons. The energy was minimized with respect to all parameters and a minimum energy -2.17522 a.u. [E (exptl) $= -2.175216$].

For treatment of (2 ^{1}S) Hylleraas and Undheim used a similar ϕ, with sine and cosine terms interchanged, so that the function was symmetric. They minimized the second lowest root of the secular equation with respect to all parameters and found a minimum energy -2.1449 a.u., compared with the experimental value -2.1460 a.u. Coolidge and James[25] with a better trial function obtained a minimum energy -2.1458 a.u.

Another method which gives excellent results for excited states of He is based on a perturbation expansion in inverse powers of the nuclear charge.[26] The Schrödinger equation is transformed by use of $1/Z$ as unit of length and Z^2 as unit of energy into

$$(H_0 + 1/Zr_{12} - E)\psi = 0,$$

$$H = -\nabla_1^2/2 - \nabla_2^2/2 - 1/r_1 - 1/r_2,$$

$$H = H_0 + H_1; \qquad H_1 = 1/Zr_{12}.$$

It is assumed that the wave function may be expanded in a similar way

$$\psi_2 = \psi_0 + \psi_1/Z + \psi_2/Z^2,$$

and in units of Z^2

$$E = E_0 + E_1/Z + E_2/Z^2.$$

In the expansions the subscripts indicate the order of perturbation. The usual basic perturbation equations may be used to obtain the energy to third order. The principal difficulty is in calculation of an accurate first order function. Sharma and Coulson used a 12-term variational trial function for this purpose, similar in form to the Hylleraas function. The results for energies of the 2 ^{1}S and 2 ^{3}S states, to the third-order are: E(singlet) $=$ -2.147273; E(triplet) $=$ -2.173968. The singlet energy lies below the experimental value but may be raised by higher order terms and other corrections which were not made. More recently, the requirements of orthogonality of the 2 ^{1}S excited state function to the 1 ^{1}S ground state function have been analyzed by Cohen and Kelly.[27] In a Hartree–Fock treatment they find that the 1s orbital should not be required to be orthogonal to the 2s orbital in the excited state function, in agreement with Sharma and Coulson. They find however that the excited state function should be explicitly required to be orthogonal to the Hartree–Fock (HF) ground state function. They consider that the "most satisfactory" HF energy is -2.14295 a.u.

The correlation energy of a particular state is usually defined as the difference between the exact nonrelativistic energy and the best HF energy for the state. Cohen and Kelly find a correlation energy of 0.00303 a.u. Sharma and Coulson find about 0.002 for the 2 ^{3}S state. The correlation energy of the ground state is discussed by Linderberg.[28]

2.2 ONE ELECTRON MODELS FOR RYDBERG TERMS

Less exact methods must be used with systems of more than two electrons. A model of a single (Rydberg) electron outside a "core" of nucleus and other electrons divides an atomic system into two parts. This model certainly is reasonable physically because of the average large distance of the Rydberg electron from the nucleus and other electrons. There are difficulties however in the determination of the wave function for the Rydberg electron by variational methods. Special constraints must be introduced to prevent collapse of the Rydberg electron function into the core. It is usually required that the Rydberg function be strictly orthogonal to the core function—a restriction that is frequently artificial and always troublesome.

There are several approximate methods for the solution of this problem, all of which assume basically that the problem can be formulated as a pseudoeigenvalue equation

$$h_{\text{eff}} \, \psi = \epsilon \psi. \tag{2-6}$$

The effective Hamiltonian, h_{eff} contains core electron operators for the Rydberg electron. The ψ are assumed to be normalized antisymemetrized product functions of the form $\psi = \psi_{\text{core}} \cdot \psi_R$. The core function may be an accurate function, such as a variationally determined function of the corresponding ion, or may be the normal state function after removal of the Rydberg electron. The sensitivity of the core function to the function ψ_R may be expected to vary in particular cases; and furthermore, ψ_{core} must change to some extent in a definite Rydberg series characterized by definite n, l, but these are small changes and are usually neglected in this approximation. It appears to be possible to vary ψ_R alone with respect to a fixed or "frozen" core. The solution of an equation such as (2–6) was shown to be possible in general by Örhn and McWeeny,[29] provided that the effective Hamiltonian, h_{eff} is properly defined. The eigenvalues ϵ, which are the orbital energies of the Rydberg functions, are just the values of the terms with reversed sign, $E_{\text{(core)}} - E_{\text{(Rydberg state)}}$, or the ionization energies of the Rydberg electron. The general treatment of Orhn and McWeeny assumed no orthogonality constraints; the orthogonality problem was avoided by proper formulation of the effective Hamiltonian. Ordinarily, the effective Hamiltonian would contain a one-electron operator for the Rydberg electron in the field of a nucleus alone, and two-electron Coulomb and exchange operators for the core orbitals. The effective Hamiltonian of Öhrn and McWeeny contained additional terms (of some complexity) that served to keep the outer electron outside of the core. The additional terms possibly are unnecessary in case the Rydberg electron is made orthogonal to the core function.

The general problem of separability of an electronic system into parts had been considered earlier by Lykos and Parr,[30] McWeeny[31] with application to separability of σ and π electrons.

The problem has been formulated in a somewhat different way by Hazi and Rice,[32] in a "pseudopotential" theory. The theory is based on the Hartree–Fock (HF) treatment of atoms and molecules. The pseudopotential consists of the HF potential of the core plus a potential that prevents collapse of a variational wave function for the Rydberg electron. In open shell configurations, which usually occur for Rydberg states, the core orbitals and outer orbitals are eigenfunctions of different one-electron operators. However, there exists a set of pseudo-wavefunctions, χ, which will satisfy an eigenvalue equation

$$(G + V_R)\chi_v = \eta_v\chi_v,$$

where G is the one-electron HF operator for the outer electron and V_R is an additional nonlocal potential. The pseudo-wave function is not assumed to be orthogonal to the core. The eigenvalues η_v are identical with eigenvalues of the operator G. Hazi and Rice derive expressions for the operator $(G + V_R)$, and apply the method to calculation of triplet states of He and Be. The core function is assumed to be the wave function of the ground state of the corresponding positive ion. As before, if the Rydberg function is orthogonal by symmetry to the core function or is made to be orthogonal, a pseudopotential is not really necessary. The orthogonality condition in itself appears to be sufficient, to prevent collapse of the Rydberg function into the core.

Öhrn and McWeeny[29] calculated $2\,^2S$, $2\,^2P$ terms of lithium and $3\,^2S$, $3\,^2P$ terms of sodium. These terms are the lowest of their respective symmetries in the two atoms. The Rydberg functions were conventional Slater orbitals and were far from orthogonal with the cores, except for the 2p function of lithium. The cores in these calculations were essentially Li^+ or Na^+ and were represented by single determinantal functions.

The results are compared with orbital energies or ionization potentials of the atoms determined by SCF procedures, and are in excellent agreement with the latter.

Similar calculations were made by Coulson and Stamper[33] for (np), ^2P terms of Li with $n = 2, 3, 4$. They compared results obtained with use of

(a) (1s) Slater orbital core, and hydrogenic (np) Rydberg function,

(b) (1s) Slater core and Slater (np) Rydberg function,

(c) (1s) Slater core with (2p) Rydberg function, orbital parameter determined by variation,

(d) SCF core (Roothan et al.[34]) with hydrogenic Rydberg functions.

The result with (c) agrees exactly with the result of Öhrn and Mc-Weeny, as would be expected. The result for (d) is still better for (2p) and both (2p) and (3p) agree closely with exponential term. They concluded that use of a hydrogenic orbital for the outer electron was justified. In general, they found that the detailed form of the core orbital was less critical than the form of the Rydberg orbital. In the case of Li, all hydrogenic (np) are also orthogonal to the core and to each other. Simple forms of (2p), (3p), and (4p) are orthogonal to the core, but are not mutually orthogonal, and so would be unsuitable for trial variational Rydberg functions, unless they were orthogonalized.

Coulson and Stamper calculated separately terms which correspond to penetration energy of the Rydberg electron ϕ_R into the core and to exchange energy of the Rydberg electron with core electrons. The penetration energy (PE) for Li is defined as

$$PE = -\langle \phi_R \mid (Z-1)/r \mid \phi_R \rangle + 2J_{1s,\,\phi_R},$$

where Z is the nuclear charge ($Z = 3$ for Li) and

$$J_{1s,\,\phi_R} = \langle 1s(i)1s(i) \mid r_{ij}^{-1} \mid \phi_R(j)\phi_R(j) \rangle.$$

The exchange energy is $-K_{1s,\,\phi_R}$

$$-K_{1s,\,\phi_R} = -\langle 1s(i)\phi_R(i) \mid r_{ij}^{-1} \mid 1s(j)\phi_R(j) \rangle.$$

The total energy of a Li Rydberg state is

$$^2E = E(^1s, \text{ion}) - \langle \phi_R \mid -\nabla^2/2 \mid \phi_R \rangle - \langle \phi_R \mid Z/r \mid \phi_R \rangle$$
$$+ 2J_{1s,\,\phi_R} - K_{1s,\,\phi_R},$$

or

$$^2E_1 - E(^1s, \text{ion}) = \left[-\langle \phi_R \mid -\nabla^2/2 \mid \phi_R \rangle - \langle \phi_R(1/r)\phi_R \rangle \right]$$
$$+ PE - K_{1s,\,\phi_R},$$

or

$$-T_{(\text{Rydberg})} = E_H + PE + \text{exchange energy}.$$

It is found that the exchange energy is larger than the penetration energy for the (np) terms of Li. This result is confirmed by more recent results by Cook and Murrell,[35] who have also reported calculations of other Rydberg series terms (2p, 3p, 4p, 3d, 4d, 4f) in both He and Li. Cook and Murrell used hydrogenic functions for the Rydberg orbital and in separate calculations on Li used optomized Slater core orbitals and SCF orbitals. Use of an SCF core gave somewhat better results in the ^2P series of lithium, particularly for the lower terms, but the differences are small. These authors state further that (unpublished) results were obtained for each series in both atoms.

In the case of Li, the exchange energy was found to be larger than the penetration energy, and the former makes a large contribution to the total quantum defect, defined as the difference in energy between the actual Rydberg term and the corresponding term of the hydrogen atom.

This result seems reasonable, because the external electron is penetrating a tightly bound core of 1s electrons. In cores with less firmly bound electrons, even with Be, the penetration energy is expected to be much larger than the exchange energy. Calculations on Be by the present writer[36] support this conclusion.

The beryllium atom is the most simple example of an atom with two valency electrons. Several independent singlet and triplet series, with variety of l values are known experimentally. These series converge to the lowest state ^{2}S of Be$^+$, derived from the configuration $(1s)^2(2s)$. The core consists thus of a tightly bound 1s shell and a (2s) valence shell electron. Experimental series n ^{1,} ^{3}S, n ^{1,} ^{3}P, and n ^{1,} ^{3}D, with $n \geqq 3$ have been observed—some rather incompletely. The lowest excited states 2 ^{1}P, 2 ^{3}P are not properly classed as Rydberg states. These states have been treated accurately.[34]

Calculations of six ^{3}S terms of Be have been made by a pseudopotential method.[32] Lin and Duncan[36] calculated terms up to $n = 10$ for ^{1}S, ^{3}S, ^{1}P, ^{3}P series. Core functions determined by Roothaan et al.[34] were used with Rydberg orbitals approximated as sums of Slater functions. The Rydberg orbitals were maintained orthogonal to the core functions and to each other. The coefficents in the linear combinations and the exponential parameters were determined by variational methods. The calculated terms are in very close agreement with experimental values of ^{1}S, ^{3}S terms, and in moderate agreement with the ^{1}P and ^{3}P terms that have been reported. Furthermore, the agreement improves with increasing n and the highest members of ^{1}S, ^{3}S are only about 20–40 cm^{-1} above the experimental values. Penetration energies were computed separately and were larger for ^{3}S than for ^{1}S. Penetration here is associated principally with the 2s unexcited electron.

Results of one-electron calculation of Rydberg terms agree quite well with terms determined by experimental spectroscopy. The qualitative agreement in energy obtained in simple cases can probably be extended, and may be helpful in correction of serious misassignments in spectra. The method probably will not be useful for small effects, such as polarization of the core by the Rydberg electron. The method is sufficient to account for most, but not all, of the Rydberg defect δ. Core polarization is discussed best by high-order perturbation theory, and part of the residual δ may be associated with such effects. This is suggested by Cook and Murrell.[35] The question is important for penetrating Rydberg orbitals and where serious departures from a central potential may occur in the neighborhood of $r = 0$.

Correlation effects between core and Rydberg electrons are also neglected in the approximate theory. These effects must account for part of the residual effect. Sinanoğlu and Tuan[37] have discussed the general problem of correlation in excited states, with particular reference to Li and Be terms.

Furthermore, there must be small changes in the core function with change of Rydberg terms of the same series. This interesting question should be studied, but extremely accurate core functions would have to be derived for each Rydberg state as a separate problem, through SCF procedures. If the core functions were found to be different, then the Rydberg orbitals would not be strictly orthogonal to each other. If this nonorthogonality were appreciable, then questions of applicability of variational procedures would appear.

Limiting forms of Rydberg orbitals with high quantum numbers have been studied in a formal way by Kotani[38]. In an atom, the core field V depends only on r, and the orbital has the central field form

$$\phi = R_{nl}(r) Y_m(\theta, \phi).$$

The radial functions R_{nl} are hydrogen-like at large values of r and so are solutions of the radial equation for the hydrogen atom (in Hartree units)

$$\frac{d^2R_{nl}}{dr^2} + \frac{2dR_{nl}}{r\,dr} + \left(2E + \frac{2}{r} - \frac{l(l+1)}{r^2}\right) R_{nl} = 0.$$

For bound states,

$$R_{nl} = (\text{const}/r) W_{\kappa,\mu}(Z),$$

where $W_{\kappa,\mu}(Z)$ is the confluent hypogeometric function with argument $Z = (-8E)^{1/2}r$, $\kappa = (-2E)^{-1/2}$, $\mu = l + \frac{1}{2}$. As E approaches zero, κ becomes very large, and the limiting form of R_{nl} is

$$R_{nl} = \text{const } (1/r^{1/2})\{\cos(\kappa\pi)J_{2\mu}(8r)^{1/2} + \sin(\kappa\pi)Y_{2\mu}(8r)^{1/2}\}, \quad (2\text{--}7)$$

J_p and Y_p are Bessel functions of the first and second kinds. The solution of (2–7) for $E = 0$ is, when r is not too large,

$$R_{(l)} = (1/r)^{1/2}\{A_l J_{2\mu}(8r)^{1/2} + B_l Y_{2\mu}(8r)^{1/2}\}.$$

From a requirement of regular behavior at the origin, we obtain from Eq. (2–7)

$$\tan(\kappa\pi) = B_l/A_l,$$

$$\kappa\pi = \tan^{-1}(B_l/A_l) + n\pi,$$

$$\kappa = 1/\pi \tan^{-1}(B_l/A_l) + n = \delta_l + n,$$

and since

$$\kappa = (-2E)^{-1/2},$$

$$E = -\tfrac{1}{2}(n + \delta_l)^2 = -\tfrac{1}{2}(n^*)^2,$$

which is the correct form for the energy of Rydberg terms.

The energy of Rydberg terms of atoms thus depends on l, through δ_l, which is constant for $E \to 0$, or large n and different Rydberg series may be distinguished by their l values. The higher members of each series are distinguished by n^* when n is large and δ_l is constant. There may be some ambiguities concerning n^* and n for lower series members, especially when δ_l is large and varies with n, which occurs with penetrating orbits of low l.

3

Rydberg Series of Diatomic Molecules

3.1 INTRODUCTION. SERIES IN H_2^+, H_2, AND HE_2

A number of factors cause Rydberg transitions in diatomic molecules to be more complicated than for similar transitions in atoms. The Rydberg electron now moves in the field of two nuclei and the internuclear distance R may assume any value from zero to infinity. This factor is simplified in practice if attention is directed first to core states in which R differs little from the equilibrium value R_e of the normal state and the lowest state of the ion. Observation of long and intense Rydberg series in absorption, corresponding to vertical excitations apparently justify this simplification. A general theoretical description of Rydberg states, as well as the forms of Rydberg orbitals will depend greatly on R, and this factor can be discussed at a later stage.

It is expected also that the angular momentum of the electronic system including the Rydberg electron will be complicated. When the core state is Σ, then the Rydberg series is characterized by the λ value of the Rydberg electron. However, this will be true only when n for the electron is small. At sufficiently low values of n, λ may be coupled to the internuclear axis and we may have several Rydberg levels corresponding to the same values of n and l, but distinguished by λ, as $3p\sigma$, $3p\pi$ and $3d\sigma$, $3d\pi$, $3d\delta$. As n for the Rydberg orbital increases, the average distance of the electron from the core is much larger than R_e, λ tends to lose its significance, and the Rydberg electron is better characterized by its l value.

There are also complications at high values of n, when the energy separation between successive Rydberg series members becomes of the order of or smaller than the energies of nuclear notion. When this occurs, the usual Born–Oppenheimer approximation for separation of electronic and nuclear motion is no longer valid. The failure of this approximation leads to changes with n in coupling of electronic and nuclear motion. The changes in the nature of coupling, as n increases, may be discussed in terms of transitions between the classical coupling cases of Hund. The application to Rydberg states of diatomic molecules has been discussed in detail by Mulliken.[6a] When n is small, the total wave function may be written as an antisymmetrized product

$$\psi = \mathbf{A}\psi_{\text{el}}\psi_{\text{nucl}}$$

$$= \mathbf{A}\phi_{\text{core}}\phi_R \cdot (\phi_{\text{vib}}\phi_{\text{rot}}),$$

which shows a Born–Oppenheimer separation. Here, $\mathbf{A}$ is an antisymmetrizing operator. The projection of l of the Rydberg electron on the internuclear axis gives λ, which is well defined and $\Sigma\lambda_i = \Lambda$. The Rydberg state conforms to Hund's case (a) for strong coupling of the total spin Σ to total orbital angular momentum Λ along the internuclear axis and to case (b) for weak coupling of S to Λ. For Σ states, the only contribution to Λ comes from λ of the Rydberg electron and its coupling to the spin is necessarily weak, and case (b) is followed.

At high values of n

$$\psi = [\phi_{core}\phi_{vib}\phi_{rot}]\phi_R ,$$

and ϕ_R is separated in the electronic core part of the total function. The Rydberg electron moves in the field of a rotating, vibrating core. In the limit, the l-vector of the Rydberg electron is no longer coupled to the internuclear axis, and Hund's case (d) is approached. Thus, λ is not defined in the limit, but l is well defined. Mulliken has discussed the complex details of the situation in which the coupling is incomplete.

As R increases continuously from its equilibrium value, characterization of Rydberg states becomes less definite. In general, several states of a diatomic molecule resulting from different configurations may give the same dissociation products at R_∞. Thus at large R, configurational interaction becomes important, when it is allowed by symmetry, and the forms of the wave functions change. The value R at which this interaction becomes important varies with the particular molecule. The question becomes exceedingly complicated when potential curves must satisfy the symmetry requirements of correlation and must avoid symmetry-forbidden crossings. Mulliken[6a] has shown that complexities occur even in the most simple diatomic molecules. There are resulting ambiguities in meaning of term values and in the effective principal quantum number $n^* = (n - \delta)$ for Rydberg states.

The simple physical concept of an R state as a core plus Rydberg electron also fails at sufficiently large values of R for a definite n because the average distance of a He electron from the core is no longer large in comparison with the dimension of the core. However, it is still useful to associate Rydberg states at large R with definite potential curves.

As R decreases from R_e toward $R = 0$, the description of Rydberg states by united atom functions becomes more valid. However, there must be changes in the description of the core so that correlation rules at $R = 0$ are consistent with noncrossing of potential curves of the same symmetry. The description at small R is particularly complicated when the Rydberg orbital is penetrating and when the core contains electrons with high values of n.

The hydrogen molecule ion H_2^+ is the simplest molecule, with a single electron moving in the field of a core. Discussion of the effect of changes in electronic functions with change of R becomes particularly interesting in this case. Although there are no experimental data on the excited states, the wave equation is separable in ellipsoidal coordinates, and can be solved exactly for any electronic state.[39] Furthermore, the potential curves can be calculated exactly at all values R from zero to infinity. This information allows a complete discussion of changes in the wave functions with R and correlation of states of the united atom with those of the separated atoms.

At R = 0, the states are those of He^+ characterized by n and l, and are described by united-atom wave functions with $Z = 2$. At slightly larger values of R, we can imagine the nucleus to be split, but n and l are still well defined. However, the electron is now in an axial field and the wave functions are those of an atom in an electrical field, as in the Stark effect. The component of the angular momentum in the direction of the axis is λ, which is quantized with values $\lambda = 0, 1, 2 \cdots$.

At R = ∞, the electron may be considered to be in the field of either atom, and is again described by the quantum numbers n and l, characteristic of the separated atoms, which, of course, are not, in general, the same as n and l for the united atom. As the nuclei approach slightly, again one separated atom and the electron are in the field of the other atom, and the axial field is restored. Thus λ retains its meaning for all values of R between zero and infinity, but n and l do not retain their significance as quantum numbers in the intermediate regions. There are, however, correlation rules to be obeyed: λ must not change and also a $g(u)$ state near R = ∞ must correlate with a $g(u)$ state near R $\geqq$ 0. In the particular case of H_2^+, because of exact separability in ellipsoidal coordinates ξ, η, and λ, (MO) near R = 0 are characterized by quantum numbers n_ξ, n_η, and λ are near R = ∞ by $n_{\xi'}$, $n_{\eta'}$, and λ defined by parabolic coordinates. Correlation between these quantum numbers is discussed by Mulliken.[6a] The usual noncrossing rule for states of the same symmetry is modified also for H_2^+.

The lowest state of H_2^+ ($^2\Sigma_g^+$) dissociates into two H atoms, $1s_a$ and $1s_b$. As the atoms approach slightly, it is well known that the system may be described by molecular orbitals.

$$\sigma_g 1s = N_g(1s_a + 1s_b),$$

$$\sigma_w 1s = N_w(1s_a - 1s_b),$$

$\sigma_g 1s$ correlates with $1s\sigma$ of the (almost) united atom, while $\sigma_u 1s$ correlates with $2p\sigma$ of the (almost) united atom. Complete correlation diagrams are found elsewhere.[6b, 40] The ground electronic state is represented well by the MO $\sigma_g 1s$ near the extreme values of R, but the 1s atomic functions in the linear combination have $Z = 2$ near $R = 0$ and $Z = 1$ and $R = \infty$ with corresponding change in the normalization constant. The wave function at intermediate values of R is not generally well represented by a single MO; a mixture of MOs is required for a better approximation. Similarly, the lowest excited state is well represented by $\sigma_u 1s$ alone only near $R = \infty$. With decreasing values of R, $\sigma_u 1s$ is mixed with other MO and its contribution to the wave function decreases so that $\sigma_u 2p$ becomes the predominant from near $R = 0$. It is of interest to examine the behavior of spectroscopic terms of H_2^+ as R changes, in comparison with atomic terms given by the formula

$$T = \frac{Z_e^2 R}{(n - \delta)^2} = \frac{Z_c^2 R}{(n^*)^2}; \qquad Z_c = \text{core charge.}$$

Accurate T values over a wide range of R have been computed by Bates $et\ al.$[39] From these data, Mulliken[6a] has computed corresponding n^* and δ values, over the range $R = 0\text{--}9$, assuming that $Z_c = 2$. The results show that ns and $np\sigma$ MOs conform well to Rydberg series. At a definite R, n^* increases slightly to a limiting value as n increases in a series. The behavior with change of R for a definite MO depends on the type. Molecular orbitals such as σ_g, π_u, δ_g, which become sums of atomic orbitals at large R, show an increase of n^* with R. The value of δ is negative for this type, and decreases steadily from zero at $R = 0$ to a limiting value of $\delta = -1$, or $n^* = 2$. For MOs such as σ_u, π_g, δ_u, whose forms are differences of atomic orbitals at large R, δ increases from zero at $R = 0$, then decreases and may become negative at large R. Terms and δ values in Rydberg series of H_2^+ change with R and the behavior can be understood on the basis of the usual formula

$$T = Z^2/2(n - \delta)^2,$$

in Hartree units. At $R = 0$, the terms are hydrogenic, $Z = 2$ and $T = 2/n^2$. As R is increased slightly, we imagine that the core is split with a change in T and introduction of a Rydberg correction δ. If

ΔT is defined as $T(\mathrm{R}) - T(\mathrm{R} = 0)$,

$$\Delta T = 2[1/(n^*)^2 - (1/n^2)] = 2[n^2 - (n^*)^2]/n^2(n^*)^2,$$

and since

$$\delta = n - n^*, \qquad \Delta T = 2\delta(n + n^*)/n^2(n^*),^2$$

or

$$\delta = n^2(n^*)^2\Delta T/2(n + n^*).$$

When $\delta \ll n$, the expression for δ reduces to

$$\delta \simeq n^3\,\Delta T/4 \simeq (n^*)^3\,\Delta T/4.$$

The change ΔT, which results from splitting the united atom core, depends on n, l, and λ. The sign of ΔT and hence of δ depends on the sign of quantity $(l^2 + l - 3\lambda^2)$ when $l > 0$. When $l = 0$, ΔT and δ are negative. We recall that a negative ΔT corresponds to an increase in energy. Orbitals that have, at $\mathrm{R} = 0$, large values in a plane bisecting the internuclear axis, such as ns, $np\pi$, will have negative ΔT and δ at small R. These are orbitals for which $\lambda = l$, with nodes in the bisecting plane at $\mathrm{R} = 0$. Orbitals such as $np\sigma$, $nd\sigma$, $nd\pi$ will have positive ΔT and δ, corresponding to a decrease in energy.

The energy change associated with splitting of united atom arises[6b] from a change of potential V. In the united atom, the nuclei are superimposed at a point c when $\mathrm{R} = 0$, and $V_0 = -2/r_c$, with r_c the distance of the electron from c. When the nuclei are moved to points a, b on the internuclear axis, at a separation $\mathrm{R} = \mathrm{R}_{ab}$, $V = -(1/r_a) - (1/r_b)$. The change in energy is thus $\langle\phi(V - V_0)\phi\rangle$, which may be positive or negative, depending on the relative magnitudes of the potential integrals, and furthermore, the energy (and δ) may change sign as R_{ab} increases from zero.

The changes of energy, terms and δ with R in H_2^+ are associated solely with core splitting. Core splitting, its effect or change of energy and δ with R, is present in all other diatomic molecules, but is not the sole cause for these changes or even a large factor in many-electron cores. Penetration of the Rydberg electron into core may be the principal cause of change of δ but it is not always possible to separate the physical factors responsible for the change. In principle, the energy associated with core splitting in H_2 can be calculated from a difference in potential $(V - V_0)$ as in H_2^+, corrected by a scale factor which is itself a function of R. In H_2 , however there is one ($1s\sigma$) core electron. There will be a singlet and a triplet Rydberg term for each configuration and a different total δ associated with each term. A change of δ and of energy of Rydberg terms with R is therefore somewhat complicated, particularly if a discussion of core splitting and penetration is to be made. The discussion is simplified if the terms, and accordingly δ are averaged for singlet and triplet terms arising from the same configuration.

At R $= 0$, δ is ascribed mostly to penetration and decreases strongly with increase of l, as in an atom. As R is increased, the energy of the term and δ changes because of core splitting, and the total δ of an observed term may be regarded as a simple sum of δ (penetration) $+$ δ (core splitting). The part of δ due to penetration is always positive, but the part due to core splitting may be positive or negative, with the result that at some value of R > 0, the observed δ may become zero. For Rydberg terms with $\lambda = l$, as for (ns) and ($np\pi$), the calculated δ (core splitting) is negative, while for ($np\sigma$) it is positive. The change of total δ with R for the H_2 Rydberg terms appears to be rather small for $\lambda = l$, at least from R $= 0$ to R $= R_e$.

At R $\gg R_o$, as in H_2^+, Rydberg states cannot, in general, be described by a single configuration, and a change in the nature of the total wave function, so that a smooth transition into a pure linear combination of atomic orbitals (LCAO) from at R $= \infty$ will occur. Mulliken concludes that ($n - \delta$) values, and hence δ, determined for experimental Rydberg terms of H_2 are probably significant out to about R $= 1.5$ R_e or about 3 Å. Beyond this configuration, interaction becomes important and the Rydberg states are of a more complicated type. Complications occur in the potential energy curves for the Rydberg states, leading in some cases to maxima between R $= R_e$ and R $= \infty$.

The preceding discussion of Rydberg states of H_2 is based on experimental term values.* Further insight may be obtained through comparison of terms calculated with approximate wave functions and assumed potentials with the experimental terms. Two calculations on the H_2 states have been reported.

---

* Recent experimental measurements on Rydberg States of H_2 have been reported but could not be included in this disscussion (see Takezawa[41]).

A perturbation treatment to the first order has been carried through by Matsen and Browne.[42] The total electronic Hamiltonian for H_2 in Hartree atomic units may be factored into terms as follows

$$H = \left(\frac{-\nabla_i^2}{2} - \frac{1}{r_{a_i}} - \frac{1}{r_{b_i}}\right) + \left(\frac{-\nabla_j^2}{2} - \frac{1}{r_j}\right) + \left(\frac{1}{r_{ij}} + \frac{1}{r_j} - \frac{1}{r_{a_j}} - \frac{1}{r_{b_j}}\right).$$

Core (H_2^+) terms   Rydberg terms   Perturbation terms

Here i refers to the core (1s) electron, j to the Rydberg electron; a, b distinguish the nuclei. The coordinate of the Rydberg electron r_j is measured from the molecular midpoint. The operators for the Rydberg electron are just those for the H atom. The zero-order functions are of the form

$$\psi^0 = [\phi_c(i) U_{nlm}(j) \pm U_{nlm}(i) \phi_c(j)] / \sqrt{2},$$

where ϕ_c is the core function, taken as the normal state function of H_2^+, U_{nlm} is the Rydberg orbital of hydrogenic form. The plus sign is associated with singlet, the minus with triplet states.

The zero-order energy is

$$E^0 = E_c - \tfrac{1}{2}n^2.$$

The second term on the right is the hydrogenic energy.

The first-order energy $E^{(1)}$ is

$$\int \psi^0 H' \psi^0 \, dV = \langle \psi^0 \mid r_{ij}^{-1} - r_j^{-1} - r_{a_j}^{-1} - r_{b_j}^{-1} \mid \psi^0 \rangle$$

$$E^{(1)} = \langle \phi_c(i)\phi_c(i) \mid r_{ij}^{-1} \mid U_{nlm}(j) U_{nlm}(j) \rangle$$
$$\pm \langle \phi_c(i) U_{nlm}(i) \mid r_{ij}^{-1} \mid \phi_c(j) U_{nlm}(j) \rangle$$
$$+ \langle U_{nlm} \mid r_j^{-1} - r_{a_j}^{-1} - r_{b_j}^{-1} \mid U_{nlm} \rangle$$

Matsen and Browne approximated ϕ_c by a five-term variationally determined function and calculated the necessary integrals for $E^{(1)}$. The total energy is

$$E^{(0)} + E^{(1)} = E_c - \tfrac{1}{2}n^2 + E^{(1)}.$$

The zero energy was taken as the exact ground state energy of H_2^+;

$E^{(0)}$ is thus an *exact* zero order energy. Rydberg state energies were calculated for the singlet and triplet series: ns, $np\sigma$, $np\pi$, $nd\sigma$, $nd\pi$, $nd\delta$, with $n = 3, 4, \ldots, 7$. The calculated results are in very good agreement with experiment, ranging from about 16% error in the $3p\sigma$ series to 1% in the $3d\sigma$ series. The error is larger in the series that show larger quantum defects.

Better agreement with experimental terms was obtained by Hazi and Rice,[43] who determined their Rydberg functions by variational methods. Their core function was determined separately by variation of a trial function consisting of a four-term linear combination of Slater functions. The Rydberg functions were orthogonalized linear combinations of Slater functions with fixed screening parameters. The coefficients in the linear combinations were determined by the linear variation method. The the functions were optimized by variation with respect to the screening parameters. The agreement is excellent except for the $(np\sigma)$, $^1\Sigma_u^+$ series, and even this series is calculated with about the same error as in the perturbation calculation of Matsen and Browne. Furthermore, the error in different series studied by Hazi and Rice is independent of the quantum defect. Hazi and Rice ascribe the dependence of error on quantum defect in the perturbation calculation to the use of hydrogenic zero-order functions, which were not used in their calculation.

Rydberg series in He_2 have been known experimentally for a long time.* The normal state of the molecule is derived from a configuration described at moderate internuclear distances R as $(\sigma_g 1s)^2(\sigma_u 1s)^2$, $^1\Sigma_g^+$. It is well known that this state is unstable. The observed excited Rydberg states show relatively small R_e. Accordingly, the excited orbitals are described frequently in united atom formulation as ns, $np\sigma$, $np\pi$, etc., and the core as $(1s\sigma)^2(2p\sigma)$, $^2\Sigma_u^+$. The $(np\pi)$, $^{1,\,3}\Pi_g$ series are most extensive. There are shorter $(ns\sigma)$ series of both multiplicities and fragmentary nd and $np\sigma$ series. Transitions connecting the stable Rydberg states occur in accessible regions of the spectrum where they may be studied with grating spectrographs of high resolving power. The resolved rotational structure has provided detailed information on coupling relations in highly excited terms of Rydberg series. The change of coupling of rotational and electronic motion

---

* For references to the older work on He_2, see the literature.[44]

with change of R has been discussed in detail by Mulliken[6b] and others.[44]

At $R = R_e$, which is remarkably constant for all the various Rydberg states, the states are (ns), $^{1, \, 3}\Sigma_u^+$; $(np\sigma)$, $^{1, \, 3}\Sigma_g^+$; $(np\pi)$, $^{1, \, 3}\Pi_u$; $(nd\sigma)$, $^{1, \, 3}\Sigma_u^+$, $(nd\pi)$, $^{1, \, 3}\Pi_u$; $(nd\sigma)$, $^{1, \, 3}\Delta_u$. When n is small the $^3\Sigma_u^+$ and $^3\Sigma_g^+$ states conform to case (b) coupling while $^3\Pi_u$, $^3\Delta_u$ conform to case (a) or (b). Of course, there is no distinction between cases (a) and (b) for singlet states. At large n there may be transition to case (d), which would be complete if the angular momentum of the excited electron were uncoupled completely from the internuclear axis. If in addition $S = 0$, the coupling is simplified further [case (d')].[45] The angular momentum of the electron is characterized simply by l and may be very weakly coupled to the angular momentum of nuclear rotation. At equilibrium distances the motion of the Rydberg electron is almost independent of rotational motion of the core. Some of the higher nd states of He_2 are examples of almost complete l uncoupling.[46, 47, 48]

The $(np\pi)$, $^{1, \, 3}\Pi$ and $(ns\sigma)$, $^{1,3}\Sigma_u^+$ series fit formulas

$$T_n = \tfrac{1}{2}(n - \delta_n)^2; \qquad \delta_n = \delta_\infty + a_n^{-1} + b_n^{-2}.$$

This writer has evaluated the parameters δ_∞, a, b for the four series from least-squares fit of the experimental data.[40] The parameters and range of n are shown in the tabulation below.

Series	δ_∞	a	b	n
$(np\pi)$, $^1\Pi$	0.032679	0.0110895	-0.012667	2–8
$(np\pi)$, $^3\Pi$	0.67141	0.021737	-0.023025	2–10
$(ns\sigma)$, $^1\Sigma_u^+$	0.078934	0.12931	-0.013453	2–4
$(ns\sigma)$, $^3\Sigma_u^+$	0.18791	-0.67498	0.22917	2–6

The error in δ_n computed is less than 0.1% in all cases except for $(9p\pi_g)$, $^3\Pi$. Possibly this experimental term is in slight error.

Term values of other He_2 states are given in Table 39 of the work by Herzberg.[40] Here the data are not sufficiently complete to justify calculation of an equation for δ_n. Many transitions in He_2 have recently been reinvestigated,[46, 47] and, in view of l uncoupling of the various states, some revisions of previous analyses have been made. A summary of the recent work is not yet available.

It may be noted that δ for the $nd\lambda$ series are small but irregular. It seems probable that some of the spectroscopic data has been interpreted incorrectly because the extent of l-uncoupling, when $l = 2$, was not fully realized when the analyses were made. The irregularities are of the order of 30 cm^{-1} at worst, and some of them could be ascribed to perturbations. However, the usual smooth behavior of δ with n does point out small irregularities by contrast and suggests that reexamination of the original data is in order.

Rydberg series have not been discovered in other molecules with less than fourteen electrons, although a search could probably be made in some cases. Most of the molecules in this range have chemical properties that make experimental study difficult. A few excited states have been found for the hydrides LiH through OH, but the states are interpreted as valency transitions, rather than Rydberg transitions. A similar situation exists for the molecules Li_2, B_2, C_2, BN, BeO, and CN. Rydberg series have been found recently in BeH, CH, and BF. These series will be discussed later in this chapter.

No absorption spectra are found for Be_2 because its ground state, like He_2, is probably unstable. No Rydberg series have been found in emission to stable excited states, as have been found for He_2.

A more favorable situation for excitation to Rydberg states arises from a strongly bonding core with a less tightly bound closed shell of electrons outside. This situation seems to be realized in N_2, CO, and O_2 and to a lesser extent in NO. All four molecules show extensive Rydberg series.

3.2. RYDBERG SERIES IN N_2

The electronic configuration of the ground state of N_2 may be formulated as

$$(1\sigma_g)^2(1\sigma_u)^2(2\sigma_g)^2(2\sigma_u)^2(1\pi_u)^4(3\sigma_g)^2, \ ^1\Sigma_g^+.$$

According to the theory of bonding and antibonding pairs, there are four bonding pairs from $(2\sigma_g)$, $(1\pi_u)$, and $(3\sigma_g)$ and one antibonding pair, $(2\sigma_u)$. The structure is thus strongly bonding. However, the electrons in $(3\sigma_g)$ are less strongly bound to the core than the others. Excitation from this orbital has relatively little effect on the core, as shown by the approximation constancy of internuclear distances in many of the excited states of N_2. Not all of the excitations are to Rydberg upper states, but all excitations are of rather high energy, and occur in the vacuum ultraviolet region. The lower energy Rydberg states occur near non-Rydberg states with some overlapping of vibrational levels, which makes a detailed analysis somewhat uncertain.

The lowest state of N_2^+ ($^2\Sigma_g^+$) results from removal of an electron from $3\sigma_g$. The Worley–Jenkins Rydberg series[49a, 49b, 50] consist of 25 members, which converge to this state of the ion. Several higher members have been found by Carroll.[51] The upper states were interpreted by Worley as $^1\Pi_u$ and the excited orbitals as $(np\pi_u)$. Worley fitted the observed series to a formula

$$\nu(\text{cm}^{-1}) = 125{,}665.8 - R/[m + 0.3450 - (0.10/m) - (0.10/m^2)]^2,$$

$$m = 2, 3, \cdots .$$

The two lowest members ($m = 2, 3$) have been identified as Σ_u^+ states which arise on orbital excitation for $(3\sigma_g)$ to $np\sigma_u$. Carroll has pointed out that there should also be neighboring $^1\Pi_u$ states that interact with the $^1\Sigma_u^+$ states and result in extensive l uncoupling in the former.

The question of assignment of the upper states in the Worley–Jenkins series has been examined recently by Ogawa and Tanaka.[52] These investigators have reexamined the spectrum of N_2 in the vacuum ultraviolet and have found additional members of the Worley–Jenkins series as well as several new Rydberg series. The lower members of the Worley–Jenkins originally were observed to be double headed, which Worley attributed to R and Q branches. If this is so, Ogawa and Tanaka argue that the separation between heads of branches should not decrease apparently to zero at large values of n, as is observed, but the separation should approach a value of 52.3 cm^{-1}, which is the calculated distance of the R head from the band origin in an hypothetical transition whose upper state has the same B_0 value as that of the $^2\Sigma_g^+$ state of N_2^+ and lower state $^1\Sigma_g^+$ of N_2. Since this is not observed, Ogawa and Tanaka believe that the two band heads belong to different Rydberg series which converge to

$^2\Sigma_g^+$ of N$_2$. They describe the two series by equations

(I): $\nu_m = 125{,}668.8 - R/(m + 0.3697 - 0.3459/m$

    $+ 0.532/m^2 - 0.96/m^4)^2$ $m = 2, 3, 4, \cdots 31$

(II): $\nu_n = 125{,}668.8 - R/(m + 0.3142 - 0.0404/m$

    $- 0.4289/m^4)^2$, $m = 2, 3, 4, \cdots 8$

Ogawa and Tanaka suggest that the series represented by (I) is formed from orbital transitions $(\sigma_g 2p) \to (np\sigma)$, while the series (II) arises from orbital transitions $(\sigma_g 2p) \to (np\pi)$. Wilkinson and Houk[53] studied rotational perturbations in the (b') $^1\Sigma_u^+$ state of N$_2$ which they attributed to a neighboring $^1\Pi_u$ state. From the perturbed positions of certain rotational lines, Ogawa and Tanaka estimate the origin of this $^1\Pi_u$ state and find it coincides with the first member of the series (II). A further and more complicated argument is based on comparison with two transitions that have upper states derived from configurations:

    $\cdots (1\pi_u)^3(3\sigma_g)^2(3p\pi)$ and $\cdots (1\pi_u)^3(3\sigma_g)^2(4p\sigma)$.

The original paper should be consulted for the details of this argument. However the values of n* = $(n - \delta)$ in the two series is almost exactly the same. For $n = 3$, $\delta_3 = 0.733$ and 0.730 for the two series, respectively. This value seems to be of the magnitude expected for $np\pi$ series; a much larger value is expected for $(np\sigma)$ series. However, the question of assignment of the upper states as $^1\Sigma_u^+$ or $^1\Pi_u$ could not be regarded as settled.

A detailed analysis of the many states of N$_2$ below 115,000 has been published very recently by Dressler.[54] About 20 levels have been interpreted as vibrational progressions of three valence states, $(b)^1\Pi_u$, $(b')^1\Sigma_u^+$ and $(d')^1\Sigma_u^-$ (or Δ_u?) and three Rydberg states $(c)^1\Pi_u$, $(c')\Sigma_u^+$ and $(o)^1\Pi_u$. There is strong perturbation between vibrational levels of the $^1\Pi_u$ (b) and (c) states, which makes it difficult to locate the experimental $v' = 0$ of the lowest $^1\Pi_u$ state. A technique for deperturbation analysis[55] was used to obtain the actual positions of the lowest $^1\Sigma_u^+$ and $^1\Pi_u$ Rydberg states.

Transitions to vibrational levels of the upper state are observed also for many of the series members. Transitions with the same v' form series converging to the corresponding v' level of $^2\Sigma_g^+$ of the ion. In addition, Huffman et al.[56] have reported absorption from various v'' levels of the normal state to various v' levels of the upper states of the Worley–Jenkins series. Population of the $v'' > 0$ levels was achieved by activating N_2 by a microwave discharge before it was pumped through the absorption chamber. In this way many additional series are observed with principal quantum numbers up to 10.

The first excited electronic state of N_2^+ is A, $^2\Pi_u$ to which several Rydberg series converge. The $^2\Pi_u$ state comes from a configuration

$$(K, K) (2\sigma_g)^2 (2\sigma_u)^2 (1\pi_u)^3 (3\sigma_g)^2$$

Three series which converge to the $v' = 0, 1, 2$ levels of $^2\Pi_u$ were analyzed first by Worley.[57] These series have been confirmed by Ogawa and Tanaka.[52] The upper states of the series are obtained by adding $ns\sigma$ to the above configuration of N_2^+ and are ascribed to $^1\Pi_u$. Four additional new series were found also each with limits about 80 cm^{-1} below the corresponding series found by Worley mentioned above. These new series are believed to have $^3\Pi_u$ upper states that are derived from the same configuration. The $^3\Pi_u$ series, however, practically merge with the long-wavelength heads $^1\Pi_u$ series at $n \geq 8$. Furthermore, the Rydberg defects are given as identical for the singlet and triplet terms.

The Hopfield Rydberg series[58] is observed in absorption farther in the ultraviolet and appears to converge to B, $^2\Sigma_u^+$ of N_2^+. The upper states in the series probably are $^1\Sigma_u^+$ and derived from a configuration.

$$(K, K) (2\sigma_g)^2 (2\sigma_u) (1\pi_u)^4 (3\sigma_g)^2 \quad [ns\sigma \quad \text{or} \quad n\delta\sigma], \; ^{1,\,3}\Sigma_u^+.$$

The series fits a formula

$$\nu = 151240 - R/(m - 0.092)^2; \qquad m = 3, 4, 5, \cdots 10.$$

If the series is $ns\sigma$, the first member is probably $4s\sigma$ and $\delta = 1.092$, corresponding to a penetrating orbital; if the series is $(nd\sigma)$, the first member probably is $3d\sigma$ and $\delta = 0.092$. Ogawa and Tanaka[52] verified this series and extended it to $m = 20$. They observed an addition series converging to $v' = 1$ of the $^2\Sigma_u^+$ state of N_2^+.

Hopfield[58] observed originally an emission series in the same region that converged to the same $v' = 0$ level of $^2\Sigma_u^+$. Ogawa and Tanaka observed a sharp absorption series, each member of which lies to the longer wavelength side of the corresponding emission series. These investigators state that the members of the emission series are shaded toward lower frequencies, whereas the new absorption series members are shaded toward higher frequencies. The two series converge to exactly the same limit, and so both are represented by the formula

$$\nu = 151231 - R/(m + 0.1405 - 0.199/m)^2; \qquad m = 3, 4, 5.$$

If this series is $(ns\sigma)$, then the first member probably is $4s\sigma$ and $\delta_4 \simeq$ 0.9. The new absorption (emission) series is sharp, while the Hopfield series is diffuse. These facts, and the reasonable δ values of 0.092 make the assignment of the original Hopfield series to $nd\sigma$ and the new series to $ns\sigma$ at least consistent and somewhat reasonable.

3.3. Rydberg Series in CO and NO

Carbon monoxide is isoelectronic with N_2, and furthermore, there is not a large difference in the nuclear charges. The electronic configurations of the normal states of the two molecules is similar. The lowest states of the ion are the same, and in the same order. The CO molecule is, of course, heteropolar, but this fact should not have a great effect on appearance of Rydberg series in CO, in about the same spectral region where the N_2 series appear. Tanaka[59, 60] observed several Rydberg series of CO in 1942–1943, and there do not appear to have been any more recent studies.

The lowest state of CO^+ is X $^2\Sigma^+$ and results from the removal of a σ electron from the ground state configuration, as is the case with N_2. Two series converge to the $v' = 0$ and $v' = 1$ levels of the $^2\Sigma^+$ state. Each series consists of nine members and fits a formula $(v' = 0)$

$$\nu = 113029 - R/(m + 0.12)^2; \qquad m = 4, 5, \cdots 12$$

No assignment of the upper states has been proposed. By analogy to N_2, the series is possibly $n\pi$ and the upper states are $^1\Pi$. If this is the case the Rydberg denominator probably should be written as $(n - 0.88)^2$ with $n = 5, \cdots 13$. No explanation is given for absence of series members below 106,576 cm^{-1}. There are a number of lower-lying transitions which may conceal the missing members.

The first excited state of CO^+ is A $^2\Pi$, and several Rydberg series were found by Tanaka to converge to $v' = 0, 1, 2, 3, 4$ vibration levels of this state. The series with $v' = 0, 1, 2$ consist of five members each. The Rydberg denominator is given[40] as $(m + 0.30)^2$, $m = 3, \cdots 7$, which may better be written as $(n - 0.7)^2$, $n = 4, 5, \cdots$ By analogy with N_2, the upper states are also $^1\Pi$ for a configuration $(\pi)^3(np\sigma)$, or Σ^+ if the series is $(n\pi)$. The δ value (0.80) appears to make the series $(n\pi)$ more probable. The convergence limit of the series is at $133,380$ cm^{-1}.

The second excited state of CO^+ is B $^2\Sigma^+$ and Tanaka found two series (β series) converged to this state, which is $158,692$ cm^{-1} above the ground state of CO. One series is sharp and the other diffuse, but both have about the same intensity. The diffuse member lies at higher frequency and is separated from the corresponding sharp member by a small frequency difference, which decreases from 393 cm^{-1} for lowest pair $(n = 4)$ to 91 cm^{-1} for $n = 8$. Tanaka suggests that the upper states of the sharp series are $^1\Sigma^+$ and $^1\Pi$ for the diffuse series. The Rydberg denominators are $(n - 0.68)^2$ and $(n - 0.62)^2$ for the sharp and $(n - 0.62)^2$ for the diffuse series, with $n = 4, 5, \cdots 8$. Transition to higher vibrational levels $v' = 1, 2, 3$ appear also in pairs of sharp and diffuse members.

The ground state of NO^+ is identical to the normal state of CO and is $^1\Sigma^+$. The additional electron in NO which is removed at the lowest ionization potential is a π orbital. The higher states of NO^+ have not been positively identified, but several Rydberg series have been found by Tanaka[61] to converge to the higher states of NO^+. No series has been discovered which converges to the lowest $^1\Sigma^+$ state of NO^+. The position of X $^1\Sigma^+$ $(NO)^+$ above X $^2\Pi$ (NO) has been located rather precisely at $74,539$ cm^{-1} from photoionization data.[62, 63] There appears to be very good evidence[64] that the Miescher–Baer bands[65] are actually due to a transition in NO^+ from a $^1\Pi$ to X $^1\Sigma$ (ground) state. The $(0, 0)$ band of this transition lies at 73084 cm^{-1}. Now Tanaka has observed a Rydberg series $[\gamma(a)]$ in NO converging to $147,759$ cm^{-1}, which is close to the difference between X $^2\Pi$ (NO) to a $^1\Pi$ state of NO^+ which is $73,084$ cm^{-1} above X $^1\Sigma$ of NO^+. However Tanaka observed a second Rydberg series $[\gamma(b)]$ that had a convergence limit at $147,417$ cm^{-1}, also close to the $^1\Pi$ level of NO^+. The separation between levels with the same n is variable in the two series but is much larger than the $^2\Pi_{3/2}$–$^2\Pi_{1/2}$ levels in NO. Note also that the separations are too small to be ascribed to a difference between singlet and triplet states.

The ground state X $^1\Sigma^+$ of NO^+ is described by the configuration

$$(1\sigma)^1(2\sigma)^2(3\sigma)^2(4\sigma)^2(1\pi)^4(5\sigma)^2$$

and lower singlet excited states of NO^+ by configurations

or $(1\pi)^4(5\sigma)(n\sigma, n\pi, n\delta, \cdots)$, $^1\Sigma^+$, $^1\Pi$, $^1\Delta$

                $(1\pi)^3(5\sigma)^2(n\sigma)$, $^1\Pi$

                $(1\pi)^3(5\sigma)^2(n\pi)$, $^1\Sigma^+$, $^1\Sigma^-$, $^1\Delta$

                $(1\pi)^3(5\sigma)^2(n\delta)$, $^1\Pi$, $^1\Phi$

There may thus be several $^1\Sigma^+$ and $^1\Pi$ states of the ion that must be considered as possibilities for convergence of the four observed series. There does not seem to be any unambiguous way except detailed rotational analysis of assigning the upper states in the α and β series. In 1942, Tanaka[61] assumed that the β series converged to a $^3\Sigma$ state of NO^+, presumably from the configuration $\cdots$ $(5\sigma)(n\sigma)$ or $\cdots$ $(1\pi)^3(n\pi)$. It is hoped that further work will lead to more certain identification of the various series.[66]

3.4. RYDBERG SERIES IN O_2

Exceptionally complete experimental results[67] on Rydberg series in O_2 have recently been published .The lower states of O_2^+ are more positively identified. The ground state of O_2 is represented by a molecular orbital configuration, with the orbitals in order of decreasing binding energy as:

$$(\sigma_g 1s)^2(\sigma_u 1s)^2(\sigma_g 2s)^2(\sigma_u 2s)^2(\sigma_g 2p)^2(\pi_u 2p)^4(\pi_g 2p)^2, X\ ^3\Sigma_g^-.$$

The lowest state of O_2^+, X $^2\Pi_g$ results from removal of an electron from $(\pi_g 2p)^2$ with an ionization energy of 97,297 cm^{-1}. Higher states of O_2 are located at:

130,800 cm^{-1} (a $^4\Pi_u$),   135,409 (A $^2\Pi_u$),   146,556 (b $^4\Sigma_g^-$),

                  163,702 (B $^2\Sigma_g^-$),   198,125 (C $^4\Sigma_u^-$).

A few terms, but no extended series, have been found[68] which converge to the three lowest states (X, a, A) of O_2^+. It is quite possible that some of the many transitions observed by Price and Collins[69] possibly may be assigned as low members of series which converge to a $^4\Pi_u$ or $A\ ^2\Pi_u$ of O_2^+ but the evidence is not convincing because of insufficient length of series and absence of any high members. The position of the $A\ ^2\Pi_u$ states of O_2^+ is known from the $(0, 0)$ band of the second negative system of O_2^+ and the lowest ionization potential (I.P.) of O_2. The $a\ ^4\Pi_u$ state is known accurately from the $(0, 0)$ band of the first negative system, which originates from $b\ ^4\Sigma_g^-$. The latter in turn is known accurately for the convergence of Rydberg series. The positions are confirmed, although with less accuracy, by measurements of photoelectron spectroscopy.[63]

Rydberg series that converge to the $^4\Sigma_g^-$ states of O_2^+ have been studied extensively by Tanaka and Takamine[70] and recently in more detail by Yoshino and Tanaka.[67] There are four Rydberg series, one strong and one weak which converge to the zero vibrational level of $b\ ^4\Sigma_g^-$ state and one strong, one weak which converge to the zero vibrational level of the $B\ ^2\Sigma_g^-$ state of O_2^+. The upper states of the four series are not identified but the intensities of all four series seems sufficiently large to interpret them as allowed transitions. The possible upper states are therefore $^3\Sigma_u^-$ and $^3\Pi_u$. The individual transitions in the sharp series are described as single headed, whereas those in the weak series are diffuse, with no well-defined heads. It might be expected from this description of the gross appearance that the upper state of the sharp transitions is $^3\Sigma_u^-$, whereas the upper state of the weak transitions is $^3\Pi_u$. If this is accepted, then the sharp series results from excitation of an electron from a $\sigma_g 2p$ orbital to $np\sigma$ and from $\sigma_g 2p$ to $np\pi$ from the diffuse series. There is a further point in support of this supposition. The Rydberg defects for both strong series are the same within experimental errors as are the defects for both weak series. However, the defects are substantially larger for the strong series, indicating greater penetration of the excited electron. This is expected if the strong series is $np\sigma$ rather than $(np\pi)$.

Additional series are observed to higher vibrational levels of both $^2\Sigma_g^-$ and $^4\Sigma_g^-$ of O_2^+. For each strong series there is usually an accompanying weak series, although the strong series extend to higher value of v'.

At higher energies, two series have been observed[71] to converge to the lowest vibrational level of the $C\ ^4\Sigma_u^-$ state of O_2^+. Series converging to the corresponding $^2\Sigma_u^-$ state have not been observed and the position of this level is not known. The upper states of the series are assigned to $^3\Sigma_u^-$, from orbital transitions $(\sigma_u 2p) \rightarrow (ns\sigma)$ and to $^3\Pi_u$, from an orbital transition $(\sigma_u 2p) \rightarrow (nd\pi)$. This assignment is consistent with the expected Rydberg defects, which are $\delta \simeq 1.2$ for $(ns\sigma)$ and $\delta \simeq 0$ for $nd\pi$. If the last is true the n values of the $(nd\pi)$ series should all be decreased by one. Additional series converging to

TABLE II

RYDBERG SERIES IN $O_2{}^a$

O_2^+ state	v'	I.P. (cm^{-1})	δ	n	Excited orbital	State
$X\ ^2\Pi_g$		97,297	—	—	—	—
	1–4	98,943	0.55	3	$(np\sigma_u)$	$^3\Pi_u$
			0.66	3	$(np\pi_u)$	$^3\Sigma_u{}^+$
			0.74	3, 4	$(np\pi_u)$	$^3\Sigma_u{}^-$
			0.72	3, 4	$(np\pi_u)$	$^1\Delta_u$
			0.72	3	$(np\pi_u)$	$^1\Sigma_u{}^+$
$a\ ^4\Pi_u$	0	130,800	1.06	3–5	$(ns\sigma_g)$	$^3\Pi_u$
	1–7	131,008	0.18	3	$(nd\sigma_g)$	$^3\Pi_u$
			0.11	3	$(nd\pi_g)$	$^3\Sigma_u{}^-$
			0.02	3	$(nd\delta_g)$	$^3\Pi_u$
$A\ ^2\Pi_u$	0	135,409	1.06	3–5	$(ns\sigma_g)$	$^3\Pi_u$
$b\ ^4\Sigma_g{}^-$	0	146,556	1.5	5–17	$(np\sigma)$	$^3\Sigma_u{}^-$
(Strong)	1	147,730	1.5	5–17		
	2	148,860	1.5	5–16		
$B\ ^2\Sigma_g{}^-$	0	163,702	1.7	5–24	$(np\sigma)$	$^3\Sigma_u{}^-$
(Strong)	1	164,811	1.7	5–20		
	2	165,885	1.7	5–19		
	3	166,907	1.7	5–17		
$B\ ^2\Sigma_g{}^-$	0	163,700	1.5	7–13	$(np\pi)$	$^3\Pi_u$
(Weak, diffuse)	1	164,810	1.5	7–15		
	2	165,890	1.5	7–15		
	3	166,900	1.45	7–11		
$C\ ^4\Sigma_u{}^-$	0	198,125	1.2		$(ns\sigma)$	$^3\Sigma_u{}^-$
	0	198,125	0.0		$(nd\pi)$	$^3\Pi_u$
	1	198,125	1.2		$(ns\sigma)$	$^3\Sigma_u{}^-$

^{a} Weak transitions accompany some members of $^3\Pi_u$ series.

$v' = 1$ of $C\,{}^4\Sigma_u{}^-$ are also reported by Codling and Madden [71] corresponding to the two $v' = 0$ series. The authors also report four or five very
weak transitions which appear to accompany some members of the
${}^3\Pi \leftarrow {}^3\Sigma$ transitions but the interpretation of these is uncertain. The
high energy transitions are usually broadened by autoionization and
future resolution of rotational structure seems very unpromising.

Information on oxygen Rydberg transition is summarized in Table
II.

3.5. Rydberg Series in Bromine and Iodine

The vacuum ultraviolet absorption spectrum of Br_2 has been investigated recently by Venkateswarlu.[72] He has found over 200 resolved transitions and has identified nine Rydberg series. Five
converge to the lowest (${}^2\Pi_{3/2}$) level at 85,165 cm^{-1} and four to the
${}^2\Pi_{1/2}$ level at 88,306 cm^{-1}. The multiplet splitting of 3141 cm^{-1} is
practically the same as in the lowest state of atomic bromine. The
series converging to 85165 cm^{-1} begin with $n = 5$ and are inter-
preted as $(np\sigma_u)$, $(np\pi_u)$, $(nf\sigma_u)$, $(nf\pi_u)$, $(nf\delta_u)$ with $\delta = 2.593, 2.422,$
2.225, 1.938, and 1.843, respectively. The last three f series may begin
with $n = 4$ and the corresponding δ reduced accordingly by one, but
the δ appear quite large nevertheless for (nf) series. Also $\delta = 2.25$
appears unusually large for a $(np\pi_u)$ series.

The four series converging to 88,306 consist of $n = 5, 6, 7$ only and
have $\delta = 2.629, 2.591, 2.446,$ and 2.416. The first two are interpreted
as $(np\sigma_u)$ with upper states Π_{1u}, Π_{0u+}. The second two are to $(np\pi_u)$
with $\Sigma^+(1u)$, $\Sigma^-(O_u{}^+, 1u)$ upper states.

The same author has studied the electronic spectrum of I_2 with very
high resolution,[73] and has reported about 200 resolved electronic transitions. There are five long Rydberg series for the ground level and one
series from the $v'' = 1$ level which converge to the $(\sigma_g)^2(\pi_u)^4(\pi_g)^3$,
${}^2\Pi_{3/2g}$ state of the ion. Similarly, there are nine shorter series which
converge to the $(\sigma_g)^2(\pi_u)^4(\pi_g)^3$, ${}^2\Pi_{1/2g}$ state of the ion.

Individual states, but not series, have been found to converge to
${}^2\Pi_{3/2}$ and ${}^2\Pi_{1/2u}$ states of the ion, derived from a configuration
$(\sigma_g)^2(\pi_u)^3(\pi_g)^4$. Other band systems which are believed to come from
cores $(\sigma_g)^2(\pi_u)^2(\pi_g)^2(\sigma_u)$, ${}^2\Sigma_u{}^+$. Other systems appear to arise from a

TABLE III

RYDBERG SERIES AND STATE OF I_2

Series designation	I.P. (cm^{-1})	δ	n	Excited orbital	State
		$(\sigma_g)^2(\pi_u)^4(\pi_g)^3$, $^2\Pi_{3/2g}$ core			
c	75,814	3.588	6–36	(npσ_u)	Π_{1u}
c	75,600	3.588	6–34	(npσ_u)	Π_{1u}
e	75,814	3.449	6–34	(npπ_u)	$\Sigma^+ (0_u{}^+)$
g	75,814	1.048	4–33	(nfσ_u)	Π_{1u}
k	75,814	0.900	4–16	(nfπ_u)	$\Sigma^+ (0_u{}^+)$
l	75,814	0.843	4–14	(nfδ_u)	Π_{1u}
		$(\sigma_g)^2(\pi_u)^4(\pi_g)^3$, $^2\Pi_{1/2g}$ core			
f	80,895	3.544	6–8	(npσ_u)	$\Pi^0{}_u{}^+$
i		3.484	6–8	(npπ_u)	$\Sigma^+ (1u)$
j		3.430	6–8	(npπ_u)	$\Sigma^- (1u, 0_u{}^+)$
O		1.139	4,5	(nfσ_u)	Π_{1u}
O		1.113	4,5	(nfσ_u)	$\Pi^0{}_u{}^+$
q		1.093	4,5	(nfπ_u)	$\Sigma^+ (1u)$
r		0.909	4,5	(nfπ_u)	$\Sigma^- (1u, 0_u{}^+)$
s'		0.814	4,5	(nfδ_u)	Π_{1u}
s'		0.768	4,5	(nfδ_u)	Π_{1u}
		$(\sigma_g)^2(\pi_u)^3(\pi_g)^4$, $^2\Pi_{3/2u}$ core			
h	87,814		6	(6sσ_g)	Π_{1u}
m			5	(5dπ_u)	Π_{1u}
p			5	(5dσ_g)	$\Sigma^+ (0_u{}^+)$
		$(\sigma_g)^2(\pi_u)^3(\pi_g)^4$, $^2\Pi_{1/2u}$ core			
t, t'			6	(6sσ_g)	$\Pi_{1u}, \Pi^0{}_u{}^+$
u, u'			5	(5dσ_g)	$\Pi_{1u}, \Pi^0{}_u{}^+$

configuration $(\sigma_g)(\pi_u)^3(\pi_g)^4(\sigma_u)^2$. Vibrational transitions appear associated with a large number of the electronic transitions.

Information on the principal Rydberg series is summarized in Table III.

3.6. SPECTRA OF HYDRIDES, EXCITED STATES

The spectra of most diatomic hydrides (AH) show only a small number of electronic states. The first example studied of a stable molecule of this type was hydrogen iodide.[74] It is obvious that many hydrides are chemically unstable and are thus unsuitable for study in absorption, although recent study of BF, BH and CH has been made and will be discussed. Many stable hydrides have very low vapor pressures and laboratory absorption studies are difficult, particularly in the vacuum ultraviolet region. The concentration of nuclear charge about the heavier atom of the hydride leads to expectation of simple theoretical descriptions. We may hope that further search will be made.

The series in HI is represented by the formula

$$\nu = 89130 - R/(m - 0.30)^2; \qquad m = 3, 4, 5, \cdots 13.$$

It is possible[75] that the (continuous) A transition, with a maximum at 48,000 cm^{-1} is the lowest member ($m = 2$) of this series. The series converges to the upper component $^2\Pi_{1/2}$ of the HI$^+$ ion. No series that converges to the lower $^2\Pi_{3/2}$ level has been found. However, analogous series to both levels have been found in CH$_3$I. The HI$^+$ core has a configuration $\ldots (\sigma)^2(\pi)^3$ and the excited states are derived from a configuration $(\sigma)^2(\pi)^3\sigma^*$, where σ^* probably (Mulliken[75]) is $nd\sigma$. The upper states are $^3\Pi$ or $^1\Pi$, and the ground state is $^1\Sigma^+$ in Russell–Sanders coupling. Since the coupling probably is rather of a (Ω_c, ω) type,[76] the transitions in the Rydberg series is better described as from a 0^+ ground state to upper states with $\Omega = 1$.

Mulliken[75] has interpreted other transitions observed by Price[74] in the vacuum ultraviolet. The B and C states of HI are suggested as first members of series from a configuration $(\sigma)^2(\pi)^3(ns\sigma)$. The separation between $C(62,320 \ cm^{-1})$ and $B(56,750)$ is about the same as between corresponding transitions in CH^{3}I. Furthermore the separation is that expected as a $^2\Pi_{1/2}$–$^2\Pi_{3/2}$ interval. Corresponding C and B transitions are observed in HCl and HBr and correspond with the intervals $^2\Pi_{1/2}$–$^2\Pi_{3/2}$ observed in spectra of HCl$^+$ and HBr$^+$. The upper states corresponding to the B and C transitions depend on the type of coupling assumed. For coupling intermediate between

(Ω_c, ω) and (Λ, S) the transitions probably are

$$^3\Pi_1 \leftarrow {}^1\Sigma^+, \quad B$$

$$^1\Pi \leftarrow {}^1\Sigma^+, \quad C$$

The electronic spectrum of BF has been studied from the near infrared to about 900 A, both in emission and absorption at very high resolution, and with reciprocal dispersion ranging from 0.09 A mm^{-1} at long wavelengths to 0.4 A mm^{-1} below 1180 A. In recent work,[77] a very large number of singlet and triplet states, except the lowest excited singlet and triplet states, have been interpreted as Rydberg states. Spectra of individual transitions are relatively free from large perturbations, with the result that they are readily analyzed.

Three principal Rydberg series have been identified which arise from the ground state to singlet Rydberg states.

(I) $\quad \nu = 89,650 - R/(n - 1.04)^2; \quad n = 3\text{–}6;$

$$(ns\sigma), {}^1\Sigma^+;$$

(II) $\quad \nu = 89,650 - R/(n - 0.66)^2; \quad n = 3\text{–}9;$

$$(np\sigma), {}^1\Sigma^+;$$

(III) $\quad \nu = 89,650 - R/(n - 0.50)^2; \quad n = 3\text{–}15;$

$$(np\Pi), {}^1\Pi.$$

There are three additional series, with the same δ values that converge to 91,330 cm^{-1}, a vibrational level of the lowest $({}^2\Sigma^+)$ state of BF$^+$.

Since the rotational structure of the many Rydberg transitions has been resolved, rotational–electronic interactions doubling of Π states was studied in detail. Calculations may be made[78] of the Λ-type splittings for pure l-states with integral values. Failure to obtain agreement with observations is attributed to mixing of (n, l) states. The disagreements, however, are not large and the effects are small. Nevertheless, the evidence is positive and extremely interesing, since such mixing of (n, l) values is predicted. The extent of Λ splitting should measure the extent of λ uncoupling, or l uncoupling for a single electron, and so measures the extent of transition between Hund case (b) and case (d). In BF, the upper Π-states definitely are intermediate between cases (b) and (d).

The spectrum of the BH radical[79] shows that the upper states of some transitions show unusually large Λ doubling and thus, large l-uncoupling, which increases with n. The doubling in the $^1\Pi$ states is caused by an interacting $^1\Sigma$ state. The l-uncoupling is larger for the $^1\Delta$ state, and actually the $^1\Delta \leftarrow {}^1\Sigma$ transition is allowed by this process. The analysis of the splittings allows the identification of three Rydberg series. Since l is a proper quantum number, the series can be considered as ns, np, and nd series with $n = 3, 4$. The δ-values, 0.9, 0.50, 0.04, respectively, confirm this assignment. All series converge to an ionization potential of 9.77 ± 0.05 eV ($78,815$ cm^{-1}).

The absorption spectrum of CH below 1370 show[80] a number of transitions that can be arranged in a Rydberg series

$$\nu = 85,850 - R/(n - 0.09)^2; \qquad n = 3, 4, 5, 6.$$

The low value of δ suggests an nd series. The complexity of the first member suggests that it is a d-complex arising by l-uncoupling from the three states (Σ, Π, Δ) of the configuration $\sigma^2(nd)$. From the ionization potential, it is possible to identify two members of a np series with $\delta = 0.74$.

3.7. THEORETICAL CALCULATIONS OF RYDBERG TERMS

Self-consistent field calculations [81–83] of lower Rydberg states have been made for N_2, CO, and BF. They showed that the wave function of the molecular ground state with the removal of one electron could be used as a core function for the wave function of Rydberg states. The Rydberg state function is thus an antisymmetrical product of ψ_{core} and ψ_R. The core orbitals were those found from an SCF solution of the ground state and differences between them for different Rydberg states were neglected. The Rydberg orbital was determined by minimization of its energy in the field of a fixed core. The Rydberg orbital energy was subtracted from E_c to obtain the energy of the molecular Rydberg state. The value E_c was taken to be the experimental value of the corresponding ionization potential for the limit of the Rydberg series, if this was known.

Lefebvre-Brion *et al.*[81] compared a core function for CO obtained from the SCF ground state wave function of CO with a normal state SCF function of CO^+. The occupied orbitals in the two cases showed little difference. A similar result was found from the core functions derived from SCF calculations on NO and independently on NO^+.

In this method, it is necessary to include Rydberg orbitals in the basis set, so that a number of extra solutions with high quantum numbers are obtained. The Rydberg state functions are linear combinations of extra or virtual orbitals of the same symmetry and the contributions of each orbital was determined by a population analysis.

The results for lower Rydberg states calculated by this method agree well with experiment. The wave functions are orthogonal with each other and with the core. The method is straightforward but tedious, even with large scale computing facilities.

The requirement of orthogonality of the Rydberg orbital to all of core orbitals appears necessary in a variational calculation. Without this requirement, there is a strong possibility that the wave function of the Rydberg electron will merely be a poor approximation to a core orbital and so "collapse" into the core. However, it is possible to add a nonlocal potential to the usual Hartree–Fock operator and obtain a pseudopotential that will prevent this collapse.

The theory of a pseudopotential has been applied recently by Rice and his co-workers to various atomic and molecular cases.[84] The essential features have been described in Chapter II. The method is of special value when an accurate core function is not known. When a self-consistent field core function for the neutral atom, or for the corresponding ion is known, the orthogonalization problem is not severe and the pseudopotential becomes less valuable. On the other hand, the method is valuable for the most simple molecules because of the nonlocal nature of the pseudopotential. However the coordinates of core electrons are referred to local atomic centers and the pseudopotential must be constructed to take this fact into account. This is done by use of different forms in various local parts of the core region and a different form outside the core region. Also, in cases where the core function is unknown, the pseudopotential method appears to offer a considerable promise.

Even when the core function is known exactly, as in the case of H_2, there are obvious computational difficulties in the calculation of

Rydberg terms. Because of the symmetry of core charge, the coordinates of Rydberg orbital should be centered at the midpoint of the internuclear axis. This leads to two- and three-center Coulomb and exchange integrals involving large values of the principal quantum number. A variational calculation[85] of the second $^1\Sigma_g^+$ state of H_2 involves only three-center Coulomb and three three-center exchange integrals. While there are several methods of evaluation of such integrals, extension to Rydberg states with higher n appears to be so tedious that other methods are preferable.

The mathematical difficulties can be greatly reduced when the coordinates of the wave function are measured from one center. An accurate single center expansion of the diatomic wave function does not appear to be realistic, except for FH and possibly OH, where the electron density is concentrated largely on one atom. However, there are no known experimental Rydberg series in the spectra of these molecules for comparison with computed results. The principal difficulty inherent in single-center expansions lies in reproducing the behavior of the total molecular wave function in regions very close to the nuclei. It is usually found that a many-term expansion is necessary for satisfactory convergence. Accuracy is particularly difficult when the nuclei have similar or equal charges.

The above problems do not arise in the use of an effective one-electron Hamiltonian for the Rydberg electron. This is possible if we assume that core is unaffected (or frozen) when the Rydberg electron is excited and finally removed at ionization. This assumption limits calculation to Rydberg states and ions with internuclear distances about equal to the normal state value. The effective Hamiltonian in this case is just the differences in operators for the ion and Rydberg state, and the expectation value for the Hamiltonian is just the negative of the corresponding term. The coordinates of the Rydberg electron may be measured from the molecular midpoint, which leads to one or two simple center-integrals for the one-electron operators. A further approximation is needed for the two-electron integrals. These depend on the core electrons and therefore on two or three centers.

In a study of one series of Rydberg terms in the nitrogen molecule, Duncan[86] found an approximate set of midpoint-centered core orbitals to replace the two centered-orbitals. It was required that the "replacement" orbitals reproduce the electron density arising from the

SCF core orbitals in the core region, defined as a sphere with a fixed radius sufficiently large to contain a large fraction of the electron density. In this work, the hybridization of atomic orbitals in the SCF orbitals was considered to be unimportant and was neglected. The large contributions of 1s electrons to the energy in regions very close to the nuclei were neglected completely, by contracting the 1s electrons into the nuclei, with a corresponding reduction of effective nuclei charge. The computations were improved in subsequent work in which the core energy was varied separately. The agreement with experimental results was surprisingly good in view of the approximations made.

In the above calculation, we assumed that the upper states of the Worley–Jenkins series behaved as $(np_x\ \Pi_u$ at an equilibrium distance about equal to the observed r_e for the normal states of N_2 and of $N_2{}^+$. The Rydberg orbitals ϕ_n were orthogonalized Slater forms:

$$\phi_n = np_x\pi_u = \sum_{k=2}^{n} C_{kn}(kp_x); \qquad n = 3, 4, 5,$$

$$(kp_x) = N_k r^{k-1} \exp(-\zeta_k r) \sin\theta \cos\phi.$$

The ϕ_n were required to the mutually orthogonal, normalized, and orthogonal to the core. These requirements are sufficient to determine completely the expansion coefficients. Each ϕ_n was varied separately to determine the various ζ_k in the Slater forms. A better but more tedious procedure would be Rydberg functions defined as linear series of basis functions ϕ_n as

$$\psi_l{}^R = \sum_{l=3}^{n} b_{ln}\phi_n$$

when n is an arbitrary upper limit of desired functions. The b_{ln} are to be determined by a linear variation method. Then the ϕ_n are improved by variation of the ζ_k. This procedure could have then been used iteratively to obtain better $\psi_l{}^R$. This procedure has actually been used in calculation of Rydberg terms of the beryllium atom.[36]

4

Rydberg Terms of Polyatomic Molecules

It might be expected that, in general, conditions would be unfavorable for observation of Rydberg series in polyatomic molecules. There are many more possibilities for continuous or predissociated spectra. Usually the ionization potentials of the various electrons will occur with closer spacing than in a diatomic molecule or atom. Generally there will be more vibrational structure associated with the numerous other electronic transitions in the region where Rydberg transitions should occur.

There is, however, a compensatory factor. Some polyatomic molecules form a closer approximation to a central field than do diatomic molecules. This situation is obvious in the ground state of nonlinear symmetrical hydrides such as water, ammonia, and the methyl radical. However, the concentration of nuclear charge at a central atom is not sufficient when equilibrium configurations of the excited states and ion are widely different from the ground state configuration. This seems to be the case in molecules such as CH_4 and SF_6. The excited states are probably Rydberg type at small interatomic distances, but in regions of the upper potential surfaces accessible for the ground state, they are best described in some other way. As a result, no Rydberg series and in fact few excited states have been found in spectra of molecules with tetrahedral or octahedral symmetry.

The binding strength of the Rydberg electron to the core is also an important factor in the origin of Rydberg series. When long series are observed, excitation of a nonbonding or weakly bonding electron is usually indicated. The higher-series members are usually accompanied by very little or no vibrational structure. This fact makes identification of the series more simple.

The less tightly bound electrons are also usually associated with series converging to the lowest ionization potential, and in polyatomic molecules these are usually the only series observed. Lower energy transitions leading to higher potentials may be observed occasionally, but the number of members in any one molecule is, in most cases, insufficient to establish the existence of a definite Rydberg series.

Any classification of molecules according to occurrence of observed Rydberg series is somewhat arbitrary. Actually, most series have been found incidently in more general studies of spectra and electronic structure excited states. Simple hydrides were among the first of classes studied, and it is appropriate to discuss them first.

4.1. RYDBERG SERIES IN HYDRIDES WITH CENTRAL ATOM

One of the earliest and possibly the most thorough study was made by Price[87] on the water molecule. He found four series that converged to the lowest ionization potential at 101,780 cm^{-1}. In all series, the electron is excited from a b_1 (x) orbital. The series are referred to as A, B (weaker) and C, D (stronger). The A-series can be interpreted as $(b_1)(npa_1)$, $^1B_1$, and the B series as $(b_1)(npb_1)$, $^1A_1$. The two rotational structure of the first member of the (npa_1) series in H_2O (and D_2O) has been resolved and analyzed by Johns.[88] Two stronger series C and D appear to merge after the first two members, and may be interpreted as $(b_1)[(n+1)s]$ $^1B_1$ and $(b_1)(nd)$. In addition, diffuse progressions at higher frequencies have been observed by Henning.[89] Probably these converge to excited states [90] of H_2O^+.

Calculations of these series by Lin and Duncan[91] lend support to these assignments. The computed terms of the A-series with Rydberg function approximated as a linear combination of np_z atomic functions leads to very good agreement with experiment. Theoretical results of similar quality are obtained for the npb_1 series with Rydberg functions approximated as a series of np_x atomic functions. Both series start with $n = 3$. When the same method is applied to C and D, we find that the C(ns) series begins with $n = 3$, instead of $n = 4$. The first member lies in the continuous absorption with a broad maximum at 60,400 cm^{-1} while the first observed member of Price's C-series has $n = 4$. This change of assignment had been suggested previously by Johns. Calculation of the D series was made with use of nd_{zz} functions and the agreement with the first two members agree well with experimental values of Price. The accuracy of the calculations is probably sufficient to distinguish clearly between the series, but the energy values are not of spectroscopic accuracy. Thus the difference between corresponding terms of the C and D series is several hundred wave numbers in error when compared with the difference between calculated terms.

The Rydberg terms of water are closely related to terms of the oxygen atom, a fact first pointed out by Price. This idea has been used as a base for perturbation calculations by LaPaglia,[92] who has computed some of the lower Rydberg states by using oxygen orbitals as zero-order functions. Calculations of that nature have also been made by Haradas and Murrell,[93] where the Rydberg levels of water were treated as perturbed levels of neon. These perturbation methods contribute substantially to our understanding of such Rydberg states, although the numerically calculated positions of the levels do not agree as well with experiment as do the calculations of Lin and Duncan.

The ionization potentials of related molecules H_2S, H_2Se, and H_2Te

are lower than in water, and the Rydberg levels lie in a more accessible spectroscopic region. A careful and complete study of H_2S, D_2S, H_2Se, D_2Se, H_2Te, and D_2Te has been made by Price et al.[94] The very small relative shifts produced by deuterium substitution prove that the series members are separate electronic transitions that are unaccompanied by vibrations. The assignments of the Rydberg transitions in H_2S are supported further by observation of series in methyl mercaptan[94] and dimethyl sulfide.[94]

All of the above-mentioned molecules show prominent series characterized by $\delta = 1.05$ (H_2O), 2.04 (H_2S), 2.05 (MeHS), 3.05 (H_2Se), and 3.95 (H_2Te). These series are most probably (ns) with $n = 4$ (H_2O), 5 (H_2S), 6 (H_2Se), and 7 (H_2Te). These series correspond to the C terms of water and are probably $[(n + 1)s]$. They are designated as E terms in the above-listed molecules. The series referred to as "A" in H_2S corresponds to Price's A (or B) series in H_2O and is characterized by $\delta = 1.57$. A similar series is found in MeSH, but apparently have not been identified in H_2Se and H_2Te. No series was identified in $(Me)_2S$.

There are two additional short series (C and D) in H_2S, each with four members. No series formulas for them are given, but the present writer has calculated the quantum defects. The values of δ fluctuate with n but average to 2.27 for the C and 2.13 for the D series. The D series is described as intense but diffuse, whereas C series is sharper and shows a rotational profile resembling the A series. From the δ values alone, it is possible that one or both may be (nd) series. Less complete similar series are found in MeSH and in H_2Se.

The CH_2 molecule[95] has a linear structure in its ground state $^3\Sigma_g^-$. The electronic configuration is $(2\sigma_g)^2(1\sigma_u)^2(1\pi_u)^2$. The lowest linear excited states come from the configurations:

$$\ldots (1\pi_u)(3d\pi_g),\ B^3\Sigma_u^-; \qquad \ldots (1\pi_u)(3d\sigma_g),\ C^3\Pi_u;$$

$$\ldots (1\pi_u)(3d\delta_g),\ D^3\Pi_u$$

The $B^3\Sigma_u^-$ is the lowest member of a Rydberg series of four members, leading to an ionization potential of 83,851 cm^{-1}. The same spectrum is shown by CHD and CD_2 and becomes sharper on deuterium substitution. The series probably does not converge to the lowest state of the ion, which possibly is bent, but to a slightly higher linear excited state. There are also Rydberg series of Π_u states but the series formula is not stated. Theoretical calculations of the lower states of CH_2 have been made by Coulson and Stamper.[33] The calculations are based on a core of Slater orbitals and a (3d) hydrogen-like function for the Rydberg orbital. The calculated molecular terms lie in the correct region but in a different order than was found experimentally by Herzberg.

The CH_3 molecule[96, 97] is planar in its normal state and apparently is planar in all of its excited states. The normal state is based on a configuration $(1a_1')^2(2a_1')^2(e')^4a_2''$, $^2A_2''$ and transitions are allowed in D_{3h} symmetry to $^2A_1'(z)$ and $^2E''(x, y)$ states. The lower excited states most probably involve excitation of the unpaired electron to (nsa_1'), $^2A_1'$; (nde''), $^2E''$; (nda_1'), $^2A_1'$. The two latter excitations lead to Rydberg series of six members each in CH_3, and seven members in CD_3, beginning with $n = 3$. Low values of δ (0.083 and 0.090) show that the two series are most probably (nd). The Rydberg series are referred to by Herzberg as $\gamma(nde'')$ and $\delta(nda_1')$. Calculations of Coulson and Stamper show[33] that $^2E''$ states lies lower than the $^2A_1'$ states derived from the splitting of (nd). The observed γ series thus is predicted to begin at longer wavelengths than the observed δ series.

A third series with a large δ (reported as 0.36) is found in CD_3 and is ascribed to nsa_1'. This series has six members, beginning with $n = 3$. It seems more probable that the effective quantum number actually is $(n - 1.36)$, and the series begins with $n = 4$. The corresponding series in CH_3 consists only of two members. The nsa_1' series begins at longer wavelengths than the other two series and its first member has been tentatively associated with the 46,205 transition in CH_3 (46,628 in CD_3). If this is true, then there is a very large penetration (an exchange) of the first member of nsa_1' in the core, so that the energy of the first excited state is lowered by about 17,000 cm^{-1} from a value predicted by the Rydberg formula.

The ammonia molecule probably is planar in most of its excited states,[98, 99] but the ground state is pyramidal and conforms to the symmetry group C_{3v}. The configuration of the ground state is $(1a_1)^2(2a_1)^2(1e)^4(3a_1)^2$, $^1A_1$. Since the relative difference in energy (about 3000 cm^{-1}) between the pyramidal and hypothetical planar form is low, the configuration in D_{3h} symmetry is $(1a_1')^2 (2a_1')^2 (e')^4$ $(a_2'')^2$, $^1A_1'$. Allowed transitions to $A_2''(z)$ and $E'(x, y)$ states are expected. The lowest excited state $^1A_2''$ at 46,136 most probably comes from excitation to $3sa_1'$ and possibly is the first member of a Rydberg series. Two transitions at higher frequencies, 69,731 and 75,205, have been fitted into a Rydberg series by Walsh and Warsop[100]

$$\nu = 82,150 - R/(n - 1.02)^2, \qquad n = 3, 4, 5.$$

All of the electronic states of ammonia are associated with vibrational progressions in the symmetrical bending frequency.[101] It is possible that higher members of Rydberg series are concealed in this structure. No further transitions are observed with which the other observed states[99] can be fitted into the above or into other series.

It is probably true that all of the excited states of ammonia are of a Rydberg type in that they are essentially atom-like, and so involve changes of an effective quantum number. However, they are probably all planar and the NH distance is changed from the normal state value. The ion is also probably planar. These factors are responsible for excitation of the vibrational structure associated even with the high-energy transitions. The fact that the intensity maxima are displaced from the origins of the transitions makes it difficult to establish the numbering in the vibrational progressions so that the (0–0) transitions cannot be located unambiguously. The observed frequency separation of vibrational transitions in the high energy spectra make it particularly difficult to locate weak members of expected Rydberg series close to the ionization potential. Actually some structure is observed above the lowest ionization potential, which further complicates the problem.[101, 102]

A number of transitions of PH_3, PD_3, AsH_3, AsD_3, SbH_3 have been found by Walsh and Warsop.[103, 104] Rydberg series in these molecules, which are analogous to the series for NH_3, probably exist. Transitions occur with very little shift relative to NH_3. The higher members of such series are probably concealed by the extensive and closely spaced vibrational structure, and also in some cases, by strong continuous absorption in the expected region. The series are not sufficiently extensive for determination of the ionization potentials, which are not known independently.

4.2. HYDROCARBONS WITH TRIPLE BONDS

Acetylene was one of the first polyatomic molecules to be studied over an extended range of the vacuum ultraviolet region. Price[105] was the first to discover two extensive series with two limits at 91,950 and 92,076 cm^{-1}. Acetylene is linear in its normal state, derived from a configuration

$$(1\sigma_g)^2(1\sigma_u)^2(2\sigma_g)^2(2\sigma_u)^2(3\sigma_g)^2(\pi_u)^4, \, {}^1\Sigma_g{}^+.$$

The excited states which are *allowed* and linear must be ${}^1\Sigma_u{}^+$ or ${}^1\Pi_u$, derived from the open shell configurations $(\pi_u)^3(n\pi_g)$ and $(\pi_u)^3(n\sigma_g)$, respectively. The ion configuration $\ldots (\pi_u)^3, {}^2\pi_u$ is presumably linear. The difference between limits of the two Rydberg series may possibly represent a splitting of the lowest ${}^2\Pi$ state of the ion. Subsequent investigations[106, 107] have confirmed the original observations of Price, and no other series have been reported.

The second member of the $(n\sigma_g)$ series at 80,166 has been studied under high resolution by Herzberg[108] and its upper state has been established as ${}^1\Pi_u$. Presumably, all the members of Price's class-II series have upper states ${}^1\Pi_u$, as Price concluded originally. Further support of this interpretation is given by the high value of $\delta = 0.95$, which is more characteristic of an (ns) than (nd).

Theoretical calculations that support this interpretation have been made by this author.[109] The experimental ionization potential was assumed and Rydberg functions expressed as linear sums of (ns) functions. The computed terms, after complete variation with respect to the linear and nonlinear parameters agree so very well with the experimental terms of the ${}^1\pi_u$ series that any other interpretation is improbable. The differences (Δ) between experimental and calculated terms decrease regularly with increase in n, for $\Delta = 336$ cm^{-1} at $n = 3$ to $\Delta = 37$ cm^{-1} at $n = 10$. This agreement of theory with the first long Rydberg series in a polyatomic molecule is indeed gratifying.

According to Price, the second series probably has ${}^1\Sigma_u{}^+$ upper states. On this assumption, the excited orbital must be $n\pi_g$, which, at small interatomic distances, may be described as $(nd\pi_g)$. Computation of the terms of this series with Rydberg series expressed as a sum of atomic nd_{xz} functions have been made also for this series.

The value of $\delta = 0.50$ seems a little high for a d-atomic series, but not unreasonably high for a polyatomic molecule.

It is of interest to compare C_2H_2 with N_2, in relation to their configurations. In the normal state, the configurations are the same, except that the order of binding energy of the $1\pi_u$ and $3\sigma_g$ orbitals are reversed. In both molecules, we can say that an electron from a "nonbonding" orbital is excited to a Rydberg orbital. In C_2H_2 it is the $1\pi_u$ orbital that is nonbonding, but in N_2, it is $3\sigma_g$. Possibly the $3\sigma_g$ orbital is more firmly bonding in C_2H_2 because of the inclusion of hydrogen orbitals.

According to Price and Walsh,[110] the spectrum of methyl acetylene (propyne) shows two Rydberg series. The series converge to ionization potentials of 91,100 and 91,240 cm^{-1}, with first members at 64,913 and 73,597 cm^{-1}, respectively. The series are:

(I) $\nu = 91,100 - R/(n - 0.96)^2$, $n = 3, \ldots, 12$;

(II) $\nu = 91,240 - R/(n - 0.52)^2$, $n = 3, \ldots, 13$.

Series (I) and (II) have been confirmed by further work by Darby and Walsh.[111] However, Namioka and Watanabe[112] report three series, and state that their analysis does not confirm series (I) and (II).

(III) $\nu = 83,570 - R/(n - 0.98)^2$, $n = 4, \ldots, 7$;

(IV) $\nu = 83,580 - R/(n - 0.57)^2$, $n = 3, \ldots, 9$;

(V) $\nu = 83,600 - R/(n - 0.33)^2$, $n = 3, \ldots, 10$.

The first member of (IV) has almost exactly the same frequency as the first member of (I), but this probably is accidental in view of the different Rydberg corrections. On the other hand, series (I) and (II) are very similar to series observed in acetylene,[105, 107] with a shift for series members of about 850 cm^{-1} relative to acetylene. Furthermore, Price and Walsh point out that their first series members are much less shifted from the values calculated from their series formula than are first members of series in butadiene and carbonyl derivatives. They attribute this fact to higher ionization potentials of acetylene derivatives, with a consequence that the lower Rydberg orbitals do not lie so much within the dimensions of the molecule.

The series limits of (III), (IV), and (V) are followed by very strong continuous absorption which coincides with the electron impact value of the ionization potential.[113, 114]

It should be pointed out that the difference in average ionization potentials corresponding to series (I, II) and series (III, IV, V) is 7590 cm⁻¹. Furthermore, all the members of (I, II) with $n > 4$ lie above the series limit of (III, IV, V). If both limits are real, the two sets of workers have clearly observed series leading to different states of the ion, or to different ions. The lower ionization should correspond to removal of an electron from a π orbital of the triple bond, as is the case of acetylene. It seems surprising that the lowest ionization potential of methyl acetylene should be lowered so greatly by methyl substitution. Perhaps it is more surprising that Price and Walsh obtained no indication of a lower ionization potential. It would appear that the continuous absorption observed by Watanabe should have appeared in the spectra of Price and Walsh, but there is no evidence reported for this, and the transitions above 83,500 cm⁻¹ appear to be sharp. Possibly, the situation here requires further reexamination, with samples of the same origin and chemical purity, followed by reanalysis of the spectrum.

Nakayama and Watanabe[107] studied 1-butyne (ethyl acetylene) and guided by the spectrum of methyl acetylene and an observed photoionization continuum were able to identify three Rydberg series. The series members are described as diffuse and the first series with $\delta = 0.92$ was considered doubtful. This series begins with $n = 5$. The two other series have $\delta = 0.55$ and 0.33 and begin with $n = 3$.

No vacuum ultraviolet spectra and no Rydberg series have been reported for dimethyl acetylene. However, Rydberg series have been reported both for $HC \equiv C - C \equiv CH_3$ (diacetylene) and for its dimethyl derivative, $H_3C - C \equiv C - C \equiv C - CH_3$. In diacetylene, Price and Walsh[115] find two series converging to the first ionization limit:

(I) $\nu = 87{,}060 - R/(n - 0.95)^2$, $n = 3, \ldots, 9$;

(II) $\nu = 87{,}020 - R/(n - 0.52)^2$, $n = 3, \ldots, 8$.

It is to be noted that the Rydberg corrections are very similar to series (I) and (II) of acetylene and of methyl acetylene, which indicates similarity of the Rydberg orbital in the various molecules.

The lowering of the series limit in diacetylene relative to acetylene is most probably due to conjugation in the former molecule.

Of more interest is a new set of transitions above the lowest ionization potential. These transitions represent Rydberg series converging to the next higher ionization potential. The series members are incomplete and the higher members have not been found. It is expected, because of resonance, that a linear combination of π orbitals of the two triple bonds will yield two $^2\Pi$ states of the ion. The lower will be $^2\Pi_g$ and the higher $^2\Pi_u$. In support of this view, Callomon[116] finds a transition between the states $^2\Pi_g$ and $^2\Pi_u$ in the spectrum of $C_4H_2{}^+$. The origin of this transition at 19,727 predicts the second ionization potential at about 107,000 cm^{-1}.

Dimethyl diacetylene (H_3C—$C\equiv C$—$C\equiv C$—CH_3) has been studied also by Price and Walsh.[110] There are many transitions in the 60,000–93,000-cm^{-1} region, but they have not been analyzed. Therefore, the first ionization potential cannot be determined for this molecule. However, a series

$$\nu = 92{,}810 - R/(n - 0.20)^2, \qquad n = 3, \ldots, 7$$

has been found. This series probably converges to the next higher ionization potential. If we assume that the two ionization potentials differ by about 19,700 cm^{-1}, as in diacetylene, the lower potential should occur at about 83,100 cm^{-1}—a reasonable value.

Cyanogen (C_2N_2) has been investigated in the vacuum ultraviolet region by Price and Walsh,[110] but no Rydberg series are positively identified. A system of diffuse bands appear to converge toward an ionization potential of about 13.8 eV. An electron impact value[117] for the potential is given as 13.57 eV.

4.3. HYDROCARBON MOLECULES WITH DOUBLE BONDS

The electronic spectrum of ethylene has been investigated rather thoroughly. The longer-wavelength portion has been discussed by others,[118,119,120] and the discussion will not be repeated here. Price and Tutte[121] have found three Rydberg series, all converging to the lowest ionization potential at 84,750 cm^{-1}. Wilkinson[120] reinvestigated the spectrum apparently only to about 77,000 cm^{-1} and high members of the series of Price and Tutte were not included. However, many new vibrational transitions associated with the lower series members were reported. The series given by Price and Tutte are

$(n\mathrm{R})\quad \nu = 84{,}750 - R/(n - 1.09)^2, \qquad n = 3, 4, \ldots, 7;$

$(n\mathrm{R}')\quad \nu = 84{,}750 - R/(n - 0.6)^2, \qquad n = 3, 4, \ldots, 7;$

$(n\mathrm{R}'')\quad \nu = 84{,}750 - R/(n - 0.3)^2, \qquad n = 3, 4, \ldots, 7.$

The distinguishing notation $(n\mathrm{R})$, $(n\mathrm{R}')$ is from Wilkinson. In its ground state, the ethylene molecule has a planar structure with a symmetry D_{2h}. The most loosely bound electron is in a doubly filled $(1b_u)$ orbital and excitation of an electron from this orbital to s, p_x, p_y, and p_z Rydberg orbitals lead, respectively, to B_{3u}, A_g, B_{1g}, and B_{2g} Rydberg states. Allowed transitions are only to B_{3u} states. Accordingly an (ns) series should appear and should be the only series of high intensity. This series presumably should be the $(n\mathrm{R})$ series, especially in view of $\delta = 1.09$. Further support for the identification of $(n\mathrm{R})$ with (ns) is obtained from a comparable series in allene, discussed subsequently.

There remains the problem of interpretation of the series with $\delta = 0.6$ and $\delta = 0.3$. Price and Tutte state that these series are weaker than the $(n\mathrm{R})$ series, but they seem too strong to be associated with the B_{1g}, B_{2g}, or A_g states from excitation to (np) orbitals. Perhaps $(n\mathrm{R}'')$ with $\delta = 0.30$ can be associated with (nd) excitation, but $\delta = 0.60$ seems too large for a series with (nd) excitation. Some evidence for a fourth $(n\mathrm{R}''')$ series has been obtained by Wilkinson from a transition at $73{,}011$ cm^{-1} ascribed to the $(0, 0)$ transition to $(4\mathrm{R}''')$. By assuming the ionization limit of the other series, a δ value of 0.94 is calculated. When this is done $(3\mathrm{R}''')$ is found to fall in region of other strong absorption, and cannot be observed experimentally.

The electronic spectrum of propylene has been observed by Price and Tutte[121] and by Samson et al.[122] The latter have found a Rydberg series

$(\mathrm{I})\quad \nu = 78{,}586 - R/(n - 0.15)^2, \qquad n = 4, 5, \ldots, 9,$

and a second series converging to a vibrational level 527 cm^{-1} above the ground state of the ion.

Price and Walsh[123] found two Rydberg series in 1, 3-butadiene

(I) $\nu = 73{,}115 - R/(n - 0.10)^2,$ $n = 4, 5, \ldots, 8;$

(II) $\nu = 73{,}066 - R/(n - 0.50)^2,$ $n = 4, 5, \ldots, 8.$

It is to be noted that the series, both in propylene and in acetylene, begin with $n = 4$. The low values 0.15 and 0.10 appear to indicate that the excited electron is in a d rather than in an s-orbital. If the latter were the case, the series (I) should begin with 5 and the corrections should be 1.15 and 1.10. The series in 1, 3-butadiene with $\delta = 0.50$ would suggest an electron in a p-orbital.

Two series are found also in each of the substituted analogous molecules[123] isoprene, β, γ-dimethylbutadiene, and chloroprene. There is a series with $\delta = 0.10$ and a series with $\delta = 0.50$ for each molecule, and each series begins with $n = 4$, although in some cases $n = 3$ may be tentatively identified. Interpretation of the series follows 1, 3-butadiene. The average ionization potentials are 71,337, 71,245, and 71,203 cm^{-1}, respectively, for the three molecules.

A large number of Rydberg series were observed in allene ($H_2C{=}C{=}CH_2$) by Sutcliffe and Walsh.[124] Series formulas for five of the nine series are given as:

(I) $\nu = 82{,}210 - R/(n - 1.06)^2,$ $n = 3, \ldots, 8;$

(II) $\nu = 82{,}190 - R/(n - 0.70)^2,$ $n = 3, \ldots, 7;$

(III) $\nu = 82{,}190 - R/(n - 0.55)^2,$ $n = 3, \ldots, 7;$

(IV) $\nu = 82{,}200 - R/(n - 0.40)^2,$ $n = 3, \ldots, 7;$

(V) $\nu = 82{,}210 - R/(n - 0.30)^2,$ $n = 3, \ldots, 6.$

In its ground state, allene conforms to the symmetry group D_{2d} and the Rydberg states probably conform to the same symmetry. On this supposition, the allowed excited states are B_2 and E. The outermost orbital in the ion is (e)^3 and a Rydberg electron in ns will give E states, in np_z will give B_2 in ($np_{x, z}$) will give E and in nd will give E, B_2 states. We may therefore expect to observe one ns series, two np series, and two nd series. Series (I) is well accounted for as the ns series. Comparison with the series nR in ethylene shows an almost identical Rydberg defect and a uniform shift of the frequencies of allene (I) with respect to the frequencies of ethylene nR. Series (II) and (III) appear to be the expected np series. Although δ for (IV) and (V) are high, these may possibly be the expected (nd) series. It appears difficult to account for the remaining four incomplete series with very few members. In particular, it appears to be difficult to account for more than one (ns) series, although Sutcliffe and Walsh present arguments to the contrary.

There are other states in allene that may be associated with twisted configurations of the molecule and, of course, states associated with higher ionization potentials.

It is convenient to consider next the halogen derivatives of ethylene. The interest here lies in the effect of conjugative effects, specifically with interactions of π electrons of the halogen substituents with the π electrons of the multiple carbon–carbon bond. It is, of course, excitation of the latter that is responsible for the observed Rydberg states of the parent hydrocarbon molecule. The electronic spectra of halogen-substituted acetylenes does not appear to have been investigated, and no Rydberg series are reported. More information is available for substituted ethylenes, but many of the series are fragmentary, and few general conclusions can be drawn. There appears to be a general lowering of the ionization potential on successive halogen substitution, possibly because of interaction of π electrons both in halogen and C=C bond and also because of mutual repulsion of lone pairs of adjacent halogen atoms.

Series with $\delta \simeq 1.0$ are found in ethylene and in all halogen-substituted ethylenes except tetrachloroethylene, where no series of any nature have been established. These series probably result from excitation to an s-like orbital localized in the C=C bond.

Series resulting from an orbital localized on a single atom are not found. Two Cl atoms attached to an atom as in 1, 1-dichloroethylene,[125] apparently are necessary. However, a series of this origin does not appear in trichloroethylene[126]; only the s series appears. Comparable series with low δ (0.05 and 0.07) are found in vinylchloride[126] and cis-dichloroethylene, respectively.[126] These series probably arise by excitation to d orbitals. The single series in trans-dichloroethylene[126] with $\delta = 0.7$ is difficult to interpret but probably is an s-series beginning with $n = 4$.

4.4. BENZENE AND BENZENE DERIVATIVES

Among the most extensively studied compounds with C=C bonds are benzene and its derivatives. Benzene has been most extensively studied. We discuss here only Rydberg transitions, since lower energy transitions have been thoroughly treated elsewhere.[118]

Price and Wood[127] were the first to identify Rydberg series in benzene. Their series were represented by the formulas

$$(I) \quad \nu_n = 74{,}590 - R/(n - 0.45)^2, \qquad n = 4, \ldots, 8;$$

$$(II) \quad \nu_n = 74{,}495 - R/(n - 0.03)^2, \qquad n = 4, \ldots, 9.$$

These authors also found analogous series in benzene-d_6, and noted isotopic shifts in the vibrational structure that accompanied some of the transitions. These authors also found a series of diffuse transitions leading to a second ionization limit at about 94,384 cm^{-1}. The first ionization limit can be identified with the $^2E_{1g}$ state of the ion. Price and Wood interpreted the second limit as ionization of a σ-electron. However El-Sayed et al.[128] assigned the limit to ionization of an (a_{2u}) π electron. No series formulas are given for two Rydberg series converging to this limit. Rydberg series associated with excitation of a σ-electron converge to an ionization limit at about 135,850 cm^{-1}.

The absorption spectrum below the first ionization potential has been reinvestigated thoroughly by Wilkinson.[129] He has found four Rydberg series: R, R′, R″, and R‴, all converging to the same limit 74,587 cm^{-1} = 9.247 eV. Series R corresponds to series (I) of Price and Wood and R′, with an adjustment in δ, corresponds to series (II).

Wilkinson's series formulas are

$$(R) \quad \nu_n = I - R/(n - 0.46), \qquad n = 3, \ldots, 10;$$

$$(R') \quad \nu_n = I - R/(n - 0.16), \qquad n = 3, \ldots, 11;$$

$$(R'') \quad \nu_n = I - R/(n - 0.11), \qquad n = 3, \ldots, 9;$$

$$(R''') \quad \nu_n = I - R/(n - 0.04), \qquad n = 3, \ldots, 9,$$

$$I = 74{,}587 \text{ cm}^{-1}.$$

The agreement of these formulas with experimental values is excellent except when $n = 3$, where departures are expected.

On assumption of D_{6h} symmetry, the allowed excited states will have symmetry $^1E_{1u}$ or $^1A_{2u}$. Since the Rydberg transitions are all strong, these are the expected states. In a π-electron-orbital approximation, the configuration of the normal state is $(a_{2u})^2(e_{1g})^4$, $^1A_{1g}$. The ion has the configuration $(a_{2u})^2(e_{1g})^3$ and its normal state is probably $^2E_{1g}$. The doublet splitting of this state is small and cannot be used to account for additional series.

Assignment of four series to definite orbital transitions is probably impossible without further information. It is extremely unlikely that rotational structure can be resolved, and we can only discuss possibilities. E_{1u} states may be obtained by excitation of an electron to (a_{1u}), (a_{2u}), or (e_{2u}) orbitals, whereas A_{2u} states are obtained only from excitation to an (e_{1u}) orbital. At small interatomic distances in a united atom approximation, (a_{1u}) behaves as $np\sigma$ while (a_{2u}) and (e_{1u}) behave as $np\pi$. Excitation to an (e_{2u}) orbital gives also E_{1u} states (as well as B_{1u} and B_{2u} states) but the united atoms form of (e_{2u}) is difficult to deduce. The presence of two nodal planes suggests that $nd\delta$ is a good approximation. We conclude tentatively that Rydberg series in benzene can result by excitation of an electron to united atom orbitals $np\sigma$, $np\pi$, and $(nd\delta)$, and we can account for three of the four series in this way. It appears reasonable from the δ values that the stronger series R and R' are, respectively, $np\pi$ and $np\sigma$. In R', δ is probably 1.16 and the series begins with $n = 4$. With its very small δ, R''' may be the $nd\delta$ series and allowed, or both R'' and R''' may actually be allowed only vibronically.

Using a pseudopotential method, Hazi and Rice[84] have made calculations of the first few members of both the R and R' series in an attempt to decide which was E_{1u} and which was A_{2u}. The agreement of calculated with experimental terms is entirely satisfactory, but does not distinguish between the two symmetries.

Toluene shows two Rydberg series with formulas[130]

(I) $\qquad \nu_n = 71{,}180 - R/(n - 0.50)^2, \qquad n = 4, \ldots, 7;$

(II) $\qquad \nu_n = 71{,}130 - R/(n - 0.95)^2, \qquad n = 4, \ldots, 9.$

It is stated that the difference in terms of benzene minus terms of toluene is almost constant for the two series (about 3400 cm^{-1}) for corresponding n and the difference is maintained at the corresponding ionization potentials. Thus (I) of toluene behaves like (I) (or R) of benzene and (II) of toluene like (II) (or R') of benzene. Single series in ethyl benzene and o-xylene,[131] which are similar to series (II) (or R') of benzene and toluene, are observed, but there is no series corresponding to series (I) (or R). If series (II) in benzene is associated with $(np\sigma)$ excitation, the charge in ionization limit would be expected to have more effect on the $(p\sigma)$ ionization than on the $(p\pi)$ ionization. Regardless of this speculation, the lowering of series limits benzene–toluene–o-xylene parallels the decrease in ionization potentials in this series.[132]

The spectrum of other xylenes in this region was also studied, but no Rydberg series were identified.

Long Rydberg series have been observed in monofluorobenzene and in benzotrifluoride.[131] The ionization limits are shifted to higher energies, probably because of change-transfer effects. For C_6H_5F the series are

(I) $\nu = 74{,}304 - R/(n - 0.50)^2,$ $n = 3, \ldots, 8;$

(II) $\nu = 74{,}203 - R/(n - 1.05)^2,$ $n = 3, \ldots, 11;$

and for $C_6H_3F_3$

(I) $\nu = 78{,}250 - R/(n - 0.50)^2,$ $n = 3, \ldots, 7;$

(II) $\nu = 78{,}120 - R/(n - 1.05)^2,$ $n = 3, \ldots, 11.$

It is to be noted that the δ-values for comparable series in all of the substituted benzenes remains rather constant, showing essentially the same orbital excitation of π-electrons of the benzene ring.

Vacuum ultraviolet spectra for numerous other benzene derivatives, such as stryrene, α-methylstyrene, phenylacetylene, phenylcyanide, o-, m-, p-fluorotoluenes, have been obtained by Walsh and co-workers[131] but the spectra are too diffuse to enable identification of definite Rydberg series.

The spectrum of pyridine has been studied by El-Sayed et al.[128, 133] They report four well-developed series. The first series converges to an ionization potential of 9.26 eV and is similar to a series in benzene. It is ascribed to excitation of a π-electron from an a_2 orbital. The third and fourth series converge to an ionization potential 11.56 eV, which is similar to the higher ionization potential of 11.48 eV observed by the same authors in benzene. In pyridine these series correspond to excitation from a π (b_1 orbital). The second converges to an ionization potential at about 10.3 eV, which is believed to result from excitation of an $N(a_1)$ orbital localized on the N atom.

Electronic spectra of pyrazine and pyrimidine were reported by El-Sayed,[133] but no Rydberg series were reported. However Parkin and Innes,[134] after thorough study of a sharp transition in 1, 4-$C_4H_4N_2$ (pyrazine) at 55,154 cm⁻¹, suggested that it was analogous to the π Rydberg in benzene at 55,881 cm⁻¹. Furthermore, two transitions at higher frequencies possibly belong to the same Rydberg series that extrapolates to a limit of 9.29 eV. Further details are not given.

Among the dienes, cyclohexadiene and pyrrole show no Rydberg series, but series are found in cyclopentadiene, furan and thiophene.[115, 135] The series in cyclopentadiene is represented by

$$\nu = 69{,}550 - R/(n - 0.72)^2, \qquad n = 5, \ldots, 8.$$

A similar series in thiophene is

$$\nu = 72{,}170 - R/(n - 0.10)^2, \qquad n = 5, \ldots, 8.$$

According to Price and Walsh,[115] there are two series in furan[136] that are very similar to two series in butadiene. Both series converge to essentially the same limit 73,050 cm⁻¹ (9.06 eV) Watanabe and Nakayama[136] determined the ionization potential by photoionization and found a value 0.17 eV lower. They accordingly reanalyzed the absorption spectra and found two series leading to an ionization potential 71,683 cm⁻¹ and one additional series in which one vibrational quantum was excited. The series are:

$$(R_1) \quad \nu = 71{,}678 - R/(n - 0.06)^2, \qquad n = 3, \ldots, 11;$$

$$(R_1 + \nu_3) \quad \nu = 72{,}747 - R/(n - 0.06)^2, \qquad n = 3, \ldots, 8;$$

$$(R_2) \quad \nu = 71{,}688 - R/(n - 0.82)^2, \qquad n = 4, \ldots, 8.$$

The vibrational structure accompanying some of the Rydberg transitions is interpreted as being associated with the C=C bonds and the electronic transitions are probably associated with excitation of a π-orbital. A series at higher frequencies

$$(R_3) \quad \nu = 80{,}229 - R/(n - 0.28)^2, \qquad n = 4, \ldots, 10,$$

possibly arises also by excitation from another π orbital. The ionization limit appears to be a little high for this assignment and may more probably represent ionization from an n-orbital on the O atom.

No attempt has been made to interpret the upper states in R_1, R_2, and R_3 of furan. In C_{2v} symmetry there are obviously many possibilities allowed by selection rules and no information on polarization of the transitions is available.

4.5. SATURATED HYDROCARBONS

No Rydberg series have been found in the saturated hydrocarbons methane[137] and ethane,[138] and are not expected in higher members of this class of molecules. However, well-established series have been found in the methyl halides by Price.[139a, 139b] The series in CH_3I are

(I) $\nu = 81,990 - R/(n - 0.20)^2$, $n = 7, 8, \ldots, 14$

(II) $\nu = 76,930 - R/(n - 0.25)^2$, $n = 7, 8, \ldots, 14$

The series limits are separated by 5060 cm^{-1}, which is interpretated as the splitting of spin components in the lowest ^2E state of CH_3I^+. The splitting of similar doublets in ClI and BrI has about the same numerical value. An almost identical pair of Rydberg series[139b] was found in C_2H_5I, with about the same value of δ, with $n = 5$–10, and with about the same doublet splitting. Similar pairs of series are found in CH_3Br ($\delta = 0.1$) and in CH_3Cl ($\delta = 0.5$).

In the case of CH_3I, it is most probable that the lowest ionization potential results from removal of an electron from a (5pπ, e) orbital,[140] localized on the iodine atom. The allowed upper states may be A_1 or E. Transitions to A_1 states require that the excited orbital be of π type, whereas transitions to E states are possible with σ- or π-type excited orbitals. This is all that can safely be concluded from molecular orbital theory. Mulliken[140] has discussed the question of possible excited orbitals through comparison of term values, qualitative intensities, and band-types of other alkyl and hydrogen halides. Mulliken considers $np\pi$ as the more probable excited orbital in the principal Rydberg series. An orbital transition 5p $\rightarrow$ np would, of course, be forbidden in an atom but may be possible in a molecule with moderately strong j–j coupling.

In the bromides and chlorides[139a] it is probable that the electron is excited also from a $p\pi$ orbital. The degree of j–j coupling is less, and the difference in ionization limits of the two series is, therefore, smaller than in CH_3I. In CH_3F, the situation is different. This is because the F–C bond is so much stronger than other halogen–carbon bonds. The least firmly held electron is in a πe bonding orbital. This conclusion is supported by a short Rydberg series of two members,[141] with a formula

$$\nu = 104{,}200 - R/(n - 1.05)^2, \qquad n = 3, 4.$$

The corresponding ionization limit (12.9 eV) is in approximate agreement with an electron impact value.[142] The series is probably ns, and the Rydberg states are ^1E.

There are two other electronic transitions which can be fitted to a formula

$$\nu = 124{,}219 - R/(n - 1.32)^2, \qquad n = 3, 4.$$

If this series is real, then the second ionization limit (15.40 eV) probably corresponds to a σa_1 bonding orbital. The individual lower transitions are rather broad, and this is characteristic of transitions of a bonding electron. This factor may obscure observation of higher members of both series.

A Rydberg series in H_2CCl_2 has been reported by Duncan and Zobel[143] with terms very similar to terms to those of H_3CCl, and converging to about 90,500 cm^{-1}, with $\delta = 0.50$. About seven terms, beginning with $n = 4$, can be identified. There is no doubt of the purity of the H_2CCl_2 that was used. However, photoionization measurements give a somewhat higher value (91,560 cm^{-1}) than the ionization limit of the series. These authors could not identify any Rydberg series in the vacuum ultraviolet spectrum of F_2CCl_2. The series in H_2CCl_2 probably is associated with excitation from an orbital localized in a C—Cl bond. No double series as in H_3CCl is expected however, since the lowest state of the $H_2CCl_2^+$ ion is nondegenerate.

4.6. ALDEHYDES AND KETONES

The prototype of this important group of molecules, formaldehyde,[94, 118, 144] has been most extensively studied. In this molecule, it seems most probable that the most easily ionized electron comes from an N, or lone-pair orbital attached to the oxygen atom. There may be exceptions in some of the halogen-substituted derivatives, but in general, the N-orbital remains most loosely bound. This assumption is supported by the restricted range of ionization potentials. Deviations usually can be explained by changes in polarization charge effects.

The lowest ionization potential in formaldehyde is 10.88 eV $(87,769 \cdot \text{cm}^{-1})$. Four Rydberg series with essentially the same series limit have been observed. In C_{2v} symmetry, the excited states are based on configurations $(b_2)(R)$. Allowed transitions are to A_1, B_1, B_2 states and so (R) must transform like b_2, a_2, or a_1, respectively.

We consider the observed series.

$$(I) \quad \nu = 87,809 - R/(n - 1.04)^2, \qquad n = 3, \ldots, 8;$$

$$(II_a, II_b) \quad \nu = 87,710 - R/(n - 0.7)^2, \qquad n = 3, \ldots, 9;$$

$$(III) \quad \nu = 87,830 - R/(n - 0.4)^2, \qquad n = 3, \ldots, 8.$$

Series (I), because of its large δ can be identified with $ns a_1$ and the resulting states are B_2. Series (II_a), and (II_b) are represented by the same formula, but actually only the higher terms form a single series. The lower members probably result from $np_y B_2$, giving A_1 states and $np_z a_1$ giving B_2 states. Series (III) may result from excitation to nd orbitals. If this is so, the states could be A_1, B_1, or B_2.

The lower electronic states of formaldehyde probably are slightly nonplanar. The lowest state (A_2) is definitely nonplanar and is forbidden. The Rydberg states have not been studied in sufficient detail to decide the extent of nonplanarity.

The spectrum of acetaldehyde[145] resembles rather closely the spectrum of formaldehyde, which shows that the excitation processes do not depend greatly on the lowering of symmetry to C_S, but involve excitation from N-orbitals to excited states modified only slightly by methyl-hydrogen substitution. Three Rydberg series are observed

$$(I) \quad \nu = 82,504 - R/(n - 0.90)^2, \qquad n = 3, \ldots, 10;$$

$$(II) \quad \nu = 82,505 - R/(n - 0.70)^2, \qquad n = 3, \ldots, 18;$$

$$(III) \quad \nu = 82,475 - R/(n - 0.20)^2, \qquad n = 3, \ldots, 10.$$

The difference between average ionization limits in the two molecules (0.654 eV) is almost exactly equal to the difference in ionization potentials determined by photoionization. The agreement of δ for the similar series of the two molecules leads to similar interpretations. However, apart from possible B_1 states arising from (nd) series (III), which go over into A'', all other excited states should be A' in the lowered symmetry.

Duncan[146] obtained the vacuum ultraviolet spectrum of acetone. From the numerous electronic states, on the basis of intensity, he selected three that fit into a Rydberg series converging to 82,767 cm^{-1} (10.26 eV). This value is higher than an ionization potential of 9.705 eV found later from photoionization by Watanabe,[147] who reanalyzed the spectrum and found a different series of eleven members converging to the photoionization value 78,280 cm^{-1}. The series formula is given as

$$\text{(III)} \quad \nu = 78{,}280 - R/(n - 0.03)^2, \qquad n = 3, \ldots, 12$$

This series is possibly nd and is apparently similar to series (III) in formaldehyde and acetaldehyde. The series may also be ns, in which case, δ is 1.03 and the series begins with $n = 4$, resembling series (I) of formaldehyde.

The spectrum of acrolein ($H_2C{=}CH{-}CHO$) shows three series converging to the same limit[148]

$$\text{(I)} \quad \nu = 81{,}460 - R/(n - 0.95)^2, \qquad n = 4, \ldots, 8;$$

$$\text{(II)} \quad \nu = 81{,}516 - R/(n - 0.68)^2, \qquad n = 3, \ldots, 12;$$

$$\text{(III)} \quad \nu = 81{,}500 - R/(n - 0.15)^2, \qquad n = 3, \ldots, 10.$$

The similarity in terms of acrolein to terms of formaldehyde and acetaldehyde and the magnitudes of the series limits confirms that it is an electron from the lone-pair (N) orbital that is removed on ionization. By analogy also the values of δ indicate that series (I) in all molecules is ns, (II), np, and (III), nd. If the single series in acetone is valid, it is probably ns also.

The above analogies may be extended to a single Rydberg series observed in ketene ($H_2C{=}C{=}O$), with the formula[121, 149]

$$\text{(I)} \quad \nu = 77{,}491 - R/(n - 1.07)^2, \qquad n = 3, \ldots, 8.$$

The series is probably (ns) on the basis of δ. However, the identification of the series by δ does not answer the more interesting and important question of the origin of orbital excitation. This question cannot be answered simply by qualitative consideration of orbital structures.

We may consider ketene in relation to formaldehyde, ethylene, and allene and ask what is the nature of the orbital in ketene from which ionization occurs. We cannot decide whether the electron comes from the C=C bond or the C=O bond or indeed whether the question has significance in ketene. It would appear from the position of lowest-frequency absorption that ketene resembles formaldehyde rather than ethylene or allene. From this qualitative view, we suppose that the electron is excited and ionization occurs from the lone-pair or N-orbital in ketene. A π-orbital can only be considered localized with respect to the C=C=O linear group.

A similar question arises with carbon suboxide, which has a linear structure in the normal state and probably also in the ion and excited Rydberg states. Probably, in this molecule, we should expect no localization whatever. Two Rydberg series have been found[150] with fragments of a third. The series converge to an ionization limit 85,500 cm^{-1} (10.60 eV), confirmed by photoionization measurement. The series are given as

(I) $\nu = 85{,}500 - R/(n - 1.00)^2, \qquad n = 3, \ldots, 10;$

(II) $\nu = 85{,}500 - R/(n - 0.76)^2, \qquad n = 3, \ldots, 7.$

Series (I) is interpreted as excitation from a π_u orbital to $n\sigma_g$ (or ns) Rydberg orbital. This excitation results in Π_u states. Series (II) may result from excitation to $(n\pi_g)$ orbitals, with resulting $^1\Sigma_u^+$ states.

The diazomethane molecule (H_2CNN) is isoelectronic with ketene and the absorption spectrum has a somewhat similar structure, but shifted to lower frequencies because of a lower ionization potential. Merer[151] has found one Rydberg series, with the formula

$$\nu = 72{,}585 - R/(n - 0.10)^2, \qquad n = 3, \ldots, 10.$$

In addition, two members of a series with $\delta = 0.67$ were identified.

The long series is (nd), or (ns) with $\delta = 1.10$, in which case, the numbering must be increased by one.

As a single example of a cyclic ketone we may mention that the vacuum ultraviolet spectrum of cyclobutanone has recently been studied by Whitlock and this writer.[152] Some of the prominent transitions may be fitted to a formula

$$\nu = 75{,}444 - R/(n - 1.155)^2, \qquad n = 4, \ldots, 8.$$

The value of δ is consistent with an ns series. It seems most probable that conjugative effects present in ketene and carbon suboxide are completely absent in cyclobutanone. The series limit represents most probably ionization from the lone-pair oxygen orbital, and if this is so, cyclobutanone resembles acetone in electronic structure. The ionization limits for the two molecules are sufficiently close to confirm this similarity.

Only one acid has been studied in the region where Rydberg series normally occur. Price and Evans[153] reported a series in formic acid (HCOOH)

$$\nu = 91{,}370 - R/(n - 0.60)^2, \qquad n = 3, \ldots, 6.$$

The authors suggest that excitation in this case is from the lone-pair orbital. The unusually high ionization potential is possibly because of the polar nature of the OH group.

4.7. ETHERS AND RELATED MOLECULES

In simple ethers, such as dimethyl ether, the lowest excited states and Rydberg transitions should also occur by orbital excitation from the lone-pair on the oxygen atom. From this view, the Rydberg terms should be similar to those of water and roughly to excited levels of the oxygen atom. The same considerations apply less exactly to cyclic compounds with oxygen such as ethylene oxide, dioxane and trimethylene oxide.

The vacuum ultraviolet absorption spectrum of dimethyl ether was studied by Hernandez[154] who identified several Rydberg series with a general formula

$$\nu = (80{,}330 + r\nu_7) - R/(n - 0.02)^2,$$

where $\nu_7 \simeq 450$ cm^{-1} and is a frequency of the ground state of the ion and $r = 0, 1, 2, 3$. The orbital nature of the transition is not interpreted, but the low value of δ points to a nd series. Trimethylene oxide was studied by the same author[155] and three series

$$\nu = (77{,}980 + r\nu_v) - R/(n - 0.05)^2$$

with $r = 0, 1, 2$ and $\nu_v = 150\text{--}200$ cm^{-1}. Here $n = 3\text{--}10$ for r_0, 3–7 for r_1, and 3–6 for r_2. Tetrahydrofuran showed two series

$$\nu = 75{,}960 - R/(n - 0.10)^2, \qquad n = 3, \ldots, 8;$$

$$\nu = 76{,}140 - R/(n - 0.10)^2, \qquad n = 3, \ldots, 6.$$

The separation of the series limits (180 cm^{-1}) is believed here to represent a vibrational frequency of the normal state of the ion.

There are two additional series

$$\nu = 77{,}580 - R/(n - 0.20)^2, \qquad n = 3, \ldots, 6;$$

$$\nu = 77{,}720 - R/(n - 0.20)^2, \qquad n = 3, \ldots, 5,$$

whose limits are separated by 140 cm^{-1}, which is interpreted as a vibrational frequency of a higher state of the ion.

The spectrum of the tetrahydropyran[155] is similar, but only one Rydberg series was identified, with the formula

$$\nu = 74{,}630 - R/(n - 0.1)^2, \qquad n = 3, \ldots, 8.$$

Many of the Rydberg series members in all-cyclic ethers are accompanied by vibrational transitions. The original papers should be consulted for details. It is to be noted that the δ-values for all of the above-mentioned cyclic ethers are about the same. The low values suggest that all of the series are nd and excitation is from the lone-pair oxygen orbital.

Ethylene oxide is discussed at this point as a cyclic ether, although there are indications that the electron may be excited from a C=C orbital rather than from the oxygen lone-pair orbital. The vacuum ultraviolet absorption spectrum was studied first by Liu and Duncan,[156] who find two Rydberg series

$$\nu = 87{,}236 - R/(n - 0.55)^2, \qquad n = 3, \ldots, 10;$$

$$\nu = 87{,}175 - R/(n - 0.95)^2, \qquad n = 4, \ldots, 9.$$

These authors compared the terms of ethylene oxide fitted to the above formulas with terms of acetylene and ethylene and found correlations to support the idea of excitation from a $C \equiv C$ bonding orbital. The relatively high ionization potential, supported by electron impact value further supported the orbital excitation.

However, Watanabe[157] subsequently found a lower value for the first ionization potential by photoionization measurements. He then reassigned the transitions to form one principal and one vibrational series, with the formulas

$$\nu = 85{,}200 - R/(n - 0.04)^2, \qquad n = 3, \ldots, 8;$$

$$\nu = [85{,}200 + (1150)] - R/(n - 0.04)^2, \qquad n = 3, \ldots, 8.$$

The shape of the ionization continuum supports the view that the second series converges to a vibrational level at 1150 cm^{-1} above the ground state of the ion. With the new assignment, the δ-values are more in harmony with the other ethers discussed above, which may indicate orbital excitation from the lone-pair oxygen orbital rather than from the ethylenic group. No conclusion may be drawn from relative ionization potentials. The value for ethylene oxide is lower than for formaldehyde, but higher than for acetone. It is, however, substantially lower than the ionization potential of ethylene.

It has recently[158] been suggested that $\delta = 1.04$ in the Rydberg series just given, but that the series begins with $n = 3$, and includes a lower-frequency transition (previously interpreted as a valency transition) as the first member of an ns series. It was further suggested that two other series are present, whose first members are present at low frequencies and form np and nd series. The whole spectrum is compared to that of acetaldehyde, where three series are found. However, no detailed analyses of the spectra are presented, and no series formulas are suggested.

A somewhat questionable series of three members in 1, 4-dioxane was reported by Hernandez and Duncan[159] with the formula

$$\nu = 76{,}750 - R/(n - 0.13)^2, \qquad n = 3, 4, 5.$$

The series limit is much higher than the photoionization value[160] (72,652 cm^{-1}). No series was identified in 1, 3-dioxane which was also investigated by the same authors.

4.8. RYDBERG SERIES IN CO_2 AND RELATED MOLECULES

In this section, we discuss the symmetrical molecules CO_2, CS_2 and the unsymmetrical molecules OCS and NNO (nitrous oxide). The ground state configuration of these molecules is linear. Although some of the lowest excited states of the molecules show a bent configuration, the ions appear to have linear structures in their normal and lower excited states. It is probable that the Rydberg states all have linear configurations. The linear configurations appear to be unusually stable and probably explain the unusually large number of long Rydberg series converging to the normal and excited states of the ions.

The remaining triatomic molecules NO_2, SO_2, O_3, and XeF_2 have C_{2v} symmetry in their normal states. There are some indications that NO_2 may become linear in higher Rydberg states and in some states of NO_2^+. Nothing is really known about the geometrical structures in excited states of SO_2 and O_3, although high-energy states of the former possess unusual stability. There are a number of other triatomic molecules that show no Rydberg series, but would be particularly interesting for further investigation. Among them would be linear C_3, N_3, BO_2 and bent NH_2 and CF_2, all of which show stable low-energy excited states.

Vacuum ultraviolet spectra of CO_2 were obtained first by Lynam[161] and by Leifson.[162] Later investigations of Henning[89] and Rathenau,[163] at very short wavelengths are very important. The whole region of Rydberg transitions has been investigated most recently by Tanaka et al.[164, 165] and by Tanaka and Ogawa.[166] They found many new series converging to vibrational levels of various electronic states of the ion. The work was carried out at high resolution and high accuracy. There is probably more detailed spectroscopic information on CO_2 than on any other polyatomic molecule. About 26 series have been identified in CO_2.

In the normal state, CO_2 is linear and, in the lowest state ($^2\Pi_g$) of the ion, it is also linear. It is probable that the linear configuration is maintained for all of the Rydberg states. The normal state of the molecule is derived from the configuration

$$\ldots (3\sigma_g)^2(2\sigma_u)^2(4\sigma_g)^2(3\sigma_u)^2(1\pi_u)^4(1\pi_g)^4, \, ^1\Sigma_g^+$$

Transitions are permitted to $^1\Sigma_u^+$ and to $^1\Pi_u$ molecular states in absorption from the normal state. The lowest state of the ion comes from the configuration $(1\pi_g)^3$, $^2\Pi_{3/2g}$. The $^2\Pi_{1/2g}$ component lies about 160 cm^{-1} higher.

Tanaka *et al.* found one series converging to the $^2\Pi_{3/2g}$ level

(1) $\nu = 111{,}060 - R/(n - 0.65)^2$, $n = 3, \ldots, 15$,

and four series converging to the $^2\Pi_{1/2g}$ level.

(2) $\nu = 111{,}240 - R/(n - 0.65)^2$, $n = 3, \ldots, 15$;

(3) $\nu = 111{,}250 - R/(n - 0.57)^2$, $n = 3, \ldots, 11$;

(4) $\nu = 111{,}250 - R/(n - 0.97)^2$, $n = 4, \ldots, 11$;

(5) $\nu = 112{,}510 - R/(n - 0.7)^2$, $n = 4, \ldots, 11$.

Series (2), (3), and (4) had been found previously by Price and Simpson[167] and were extended to higher members. Series (5) converges to a vibrational level (1260 cm^{-1}) of $^2\Pi_{1/2g}$.

The next higher state of CO_2^+ is $^2\Pi_{3/2u}$ and many series have been found[165, 166] to converge to both components of this level. This principal series as $(v' = 0)$ converging to $^2\Pi_{3/2u}$ is represented by

(6) $\nu = 139{,}634 - R/[(n - 0.044) - (0.34/n)]^2$.

There are four additional long series converging to vibrational levels $v' = 1, 2, 3, 4$ with a vibrational quantum about 1130 cm^{-1}. There are also two fragmentary series to $v' = 5$, and $v' = 6$.

The principal series converging to $^2\Pi_{1/2}$ is

(7) $\nu = 139{,}726 - R/(n - 0.6)^2$, $n = 3, \ldots, 6$,

and four other series converging to 140,854, 141,974, 143,090, and 144,220 cm^{-1}, corresponding to $v' = 1, 2, 3, 4$ with about the same vibrational quantum of 1130 cm^{-1}. There are also less complete series to $v' = 5, 6, 7$.

Rydberg series to the $^1\Sigma_u^+$ were found first by Henning.[89] The series have been reinvestigated by Tanaka and Ogawa.[166] The series formulas are

(8) $\nu = 145,800 - R/(n - 0.90)^2$, $n = 4, \ldots, 19$;

(9) $\nu = 145,800 - R/(n - 0.30)^2$, $n = 3, \ldots, 18$;

(10) $\nu = 147,050 - R/(n - 0.10)^2$, $n = 8, \ldots, 11$.

Series (10) converges to a vibrational level 1250 cm^{-1} above the zero level 145,800 cm^{-1} of Σ_u^+.

In addition, Tanaka et al.[164, 165] found three series that converge to the $^2\Sigma_g^+$ state of CO_2^+

(11) $\nu = 156,350 - R/(n - 0.71)^2$, $n = 4, \ldots, 9$;

(12) $\nu = 156,400 - R/(n - 0.05)^2$, $n = 3, \ldots, 8$;

(13) $\nu = 156,410 - R/(n - 0.56)^2$, $n = 4, \ldots, 8$.

It is unfortunate that the rotational structure of none of the series members in CO_2 has been resolved and analyzed. Therefore, there seems little point in speculation on the orbital transitions involved. Absorption coefficient measurements[168, 169] do not give definite information on this point.

Experimental data on Rydberg series are quite extensive also for carbon disulfide.[164, 170] Here there is even less hope in resolution of rotational structure of the Rydberg states although a fairly complete rotational analysis of several bands in the lowest electronic transition has been accomplished. In some of the lower states the molecule is bent, but in the Rydberg states the molecule is linear as in the ground state. The ion is probably linear also.

As in CO_2, the lowest state of the ion is $^2\Pi_{3/2g}$ and the splitting of the components[169] is 440 cm^{-1}.

Price and Simpson[170] found two strong series which were confirmed and extended [171]

(1) $\nu = 81,299 - R/(n - 0.44)^2$, $n = 4, \ldots, 19$;

(2) $\nu = 81,735 - R/(n - 0.46)^2$, $n = 4, \ldots, 19$.

Series (1) converges to $^2\Pi_{3/2g}$ and (2) to $^2\Pi_{1/2g}$ of CS_2^+. Tanaka et al.[171] found additional series converging to the excited states of CS_2^+. In the tabulation to be given, (3) and (4) converge to the $^2\Pi_u$ level,

(5), (6) and (7) to the $^2\Sigma_u{}^+$ level and (8) possibly to the $^2\Sigma_g{}^+$ level. Only $^2\Pi_g$ and $^2\Sigma_u{}^+$ levels of $CS_2{}^+$ are known from direct study of the spectra of the ion. However, the analogy with $CO_2{}^+$ is strong; the $^2\Pi_u$ and $^2\Sigma_g{}^+$ states should exist and series converging to them are expected. The series converging to the excited levels of $CS_2{}^+$ are

(3) $\nu = 116{,}760 - R/(n - 0.94)^2,$ $n = 3, \ldots, 14;$

(4) $\nu = 116{,}760 - R/(n - 0.58)^2,$ $n = 3, \ldots, 8;$

(5) $\nu = 130{,}540 - R/(n - 0.81)^2,$ $n = 4, \ldots, 13;$

(6) $\nu = 130{,}600 - R/(n - 0.10)^2,$ $n = 3, \ldots, 9;$

(7) $\nu = 130{,}610 - R/(n - 0.48)^2,$ $n = 3, \ldots, 7;$

(8) $\nu = 157{,}390 - R/(n - 0.95)^2,$ $n = 4, \ldots, 11;$

The vacuum ultraviolet spectrum of COS has been studied by Price and Simpson[167] and by Tanaka et al.[164, 171] They were unable to locate the lowest ($^2\Pi$) level of the ion conclusively, but they suggested two incomplete series that probably converge to this level.

(1) $\nu = 90{,}610 - R/(n - 0.38)^2,$ $n = 5, 6, (7), 8, \ldots, 11;$

(2) $\nu = 90{,}740 - R/(n - 0.44)^2,$ $n = 6, (7), 8, \ldots, 11.$

The lower members of the series could not be identified. Additional series found are:

(3) $\nu = 129{,}400 - R/(n - 0.68)^2,$ $n = 3, \ldots, 13;$

(4) $\nu = 129{,}400 - R/(n - 0.49)^2,$ $n = 3, \ldots, 8;$

(5) $\nu = 129{,}390 - R/(n - 0.12)^2,$ $n = 3, \ldots, 9;$

(6) $\nu = 129{,}330 - R/(n - 0.96)^2,$ $n = 4, \ldots, 8;$

(7) $\nu = 144{,}680 - R/(n - 0.90)^2,$ $n = 5, \ldots, 10.$

Series (3), (4), (5), and (6) probably converge to the lowest $^2\Sigma$ state of COS$^+$ (corresponding to $^2\Sigma_u{}^+$ of $CO_2{}^+$), whereas (7) converges to a higher $^2\Sigma$ state (corresponding to $^2\Sigma_g{}^+$ of $CO_2{}^+$).

The vacuum ultraviolet spectrum of nitrous oxide, NNO, was investigated first by Duncan,[172] who fitted four transitions to a series formula

$$\nu = 102{,}567 - R/(n - 0.92)^2, \qquad n = 5, 6, 7, 8.$$

In addition, two continuous regions were believed to be $n = 3, 4$ of this series. Zelikoff $et\ al.$[173] remeasured the spectra and found $n = 4$ fitted the series formula better than before and also that the continuous region corresponding to $n = 3$ contained several weak diffuse bands and one sharp band. The latter corresponded to $n = 3$ in the formula. However, the ionization limit was found by photoionization[132] to be 104,064 cm^{-1}, or 0.18 eV higher than the spectroscopic value. The spectrum has been reinvestigated recently[171] and two series converging to the photoionization value were found. The series formulas are

(1) $\nu = 104{,}000 - R/(n - 0.60)^2, \qquad n = 3, \ldots, 13;$

(2) $\nu = 103{,}300 - R/(n - 0.68)^2, \qquad n = 3, \ldots, 9.$

The δ values in these series are very irregular, varying in the first series from 0.80 to 0.43 and in the second from 0.55 to 0.93. The difference in ionization limits is ascribed to the doublet separation in the $^2\Pi$ ground state of N_2O^+.

Two series are found to converge to the next higher ionization potential with the formulas

(3) $\nu = 132{,}210 - R/(n - 0.00)^2; \qquad n = 2, \ldots, 11;$

or $(n - 1.00)^2; \qquad n = 3, \ldots, 12;$

(4) $\nu = 132{,}250 - R/(n - 0.22)^2; \qquad n = 3, \ldots, 8.$

Also there is a vibrational series associated with a ground state frequency (1288.5 cm^{-1}) of N_2O^+ with a formula

(5) $\nu = 133{,}490 - R/(n - 0.11)^2, \qquad n = 8, \ldots, 11.$

The series limits of (3), (4), and (5) are believed to correspond to the $^2\Sigma_u^+$ state of CO_2^+, to which the Henning series in CO_2 converge.

A third ionization limit is observed at 162,200 (20.10 eV) to which four series converge, with formulas

(6) $\quad \nu = 162{,}130 - R/(n - 0.31)^2, \qquad n = 3, \ldots, 11;$

(7) $\quad \nu = 162{,}200 - R/(n - 0.06)^2, \qquad n = 3, \ldots, 11;$

(8) $\quad \nu = 162{,}200 - R/(n - 0.68)^2, \qquad n = 3, \ldots, 6;$

(9) $\quad \nu = 162{,}160 - R/(n - 0.58)^2, \qquad n = 3, \ldots, 6.$

No interpretation in terms of orbital transitions has been given for any of the individual series. Comparison of the various ionization potentials of the molecules CO_2, CS_2, COS, and N_2O has been made and discussed.[171]

The spectrum of nitrogen dioxide in the vacuum ultraviolet was studied first by Price and Simpson.[174] Two prominent transitions, each accompanied by vibrational members, were fitted to a series formula

$$\nu = 99{,}500 - R/(n - 0.75)^2, \qquad n = 3, 4, (5, 6).$$

The members corresponding to $n = 5, 6$ were observed but not accurately measured. The series formula was subsequently modified by Nakayama et al.[175] to

$$\nu = 93{,}695 - R/(n - 0.79)^2, \qquad n = (4), 5, \ldots, 9.$$

Tanaka and Jursa[176] reinvestigated the region of the Rydberg series and expressed some doubt about the existence of any series leading to the lowest ionization potential. This is found by photoionization measurements as 78,890 cm^{-1}, confirmed by an indirect electron-impact value. The failure to establish a series leading to the lowest ionization potential results from evidence that the lowest state of NO_2^+ is linear, and the higher Rydberg states are linear also. The normal state is bent and thus the Rydberg transitions have low Franck–Condon factors.

The photoionization value 78,895 cm^{-1} is associated with ionization from a $4a_1$ orbital of NO_2. The next most easily ionized electron comes from a $(3b_2)$ or a $(1a_1)$ orbital and this ionization may be associated with the series limits 99,000 to 93,000 cm^{-1}.

Four series have been observed[176] which are assigned to ionization from a $(2a_1)$ orbital. The series are:

(1) $\nu = 152,100 - R/(n - 0.15)^2$, $n = 2, \ldots, 8$;

(2) $\nu = 152,290 - R/(n - 0.67)^2$, $n = 3, \ldots, 7$;

(3) $\nu = 152,160 - R/(n - 0.03)^2$, $n = 2, \ldots, 4$;

(4) $\nu = 152,390 - R/(n - 0.03)^2$, $n = 2, \ldots, 4$.

It is quite possible that δ in series (3) and (4) may actually be 1.03 and the n values should be increased by 1. Series (3) and (4) also may not be independent, but instead, (4) may be a vibrational series separated from (3) by about 230 cm^{-1}. Actually, members of series (3) and (4), except for the first members, consist of doublets with a separation of about 300–400 cm^{-1}. This feature greatly complicates the analysis. There is an additional potential at about 112,777 cm^{-1} (13.98 eV) that is difficult to assign to any orbital ionization. There is, correspondingly, no observed spectroscopic series converging at 112,777 cm^{-1}.

The spectrum of SO_2 in the vacuum ultraviolet was first investigated by Price and Simpson.[170] They were not able to identify any Rydberg series among the numerous electronic transitions, but suggested that some of the groups of transitions approached an estimated ionization potential at about 97,200 cm^{-1} (12.05 eV). The spectrum was further investigated by Golomb et al.[177] using photoelectric methods. A reinterpretation based on a photoionization determination of the lowest ionization potential (99,547 cm^{-1}, 12.34 eV) was given. The Rydberg series found were

(R₁) $\nu = 99,500 - R/(n - 0.02)^2$, $n = 3, \ldots, 9$,

and two series of the same length leading to vibration levels of the lowest state of SO_2^+, 100,400 and 101,300 cm^{-1} above the normal state of the molecule. The frequency 900 cm^{-1} in the lowest state of SO_2^+ appears reasonable. As given, the series would be interpreted as nd, but it may be ns, with $\delta = 1.02$, and if so, the series numbering should be increased by one.

The vacuum ultraviolet spectrum[174, 178, 179] of ozone (O_3) is diffuse with six partly overlapping continua. No Rydberg series have been identified.

The vacuum ultraviolet spectrum of linear XeF_2 has been studied and shows a number of absorption transitions. Wilson et al.[180] have arranged four transitions in Rydberg series, with formulas

$$\nu = 92{,}000 - R/(n - 0.98)^2, \qquad n = 3, 4;$$

$$\nu = 98{,}000 - R/(n - 0.98)^2, \qquad n = 3, 4.$$

The difference in ionization limits (0.745 eV) is in excellent agreement with the value 0.75 for the spin–orbit coupling in atomic xenon. The high ionization potential is interpreted to indicate some π bonding. The same authors[181] have studied XeF_4 in the vacuum ultraviolet. Several transitions, but no Rydberg series, were found. The types of transitions are discussed briefly.

The present survey of Rydberg series in polyatomic molecules is necessarily incomplete, because of current publications in the subject, and indications of investigations to be published. Possibly some of the most recent work has been neglected, particularly recent copies of journals have not been received promptly. This writer can only hope that no fundamentally significant research has been missed, and if so, apologies to the authors are in order.

The most recent work points to the need of reinvestigation of many important molecules at higher frequencies, with improved continuous background for absorption and especially with greatly improved resolution. Even though complete resolution of rotational structure of some molecules cannot be expected with the largest modern instruments, more attention should be given to band profiles, in hope of more certain identification of excited states.

APPENDIX: Data on Rydberg Series in Polyatomic Molecules

This appendix consists of essential data, to April 1970, on Rydberg series of polyatomic molecules. The order in which compounds are listed follows the order of discussion in Chapter 4. Symmetry species of states in the series as well as orbital excitation are recorded when they are known.

DATA ON RYDBERG SERIES IN POLYATOMIC MOLECULES

Chemical Symbol	Name of Species	Ionization Potential (cm^{-1})	δ	Range of n	Upper States excited orbital	References
H$_2$O(D$_2$O)		101,780	0.70	3–5	$^1B_1$, (np_z)	87, 88
			0.70	3–5	$^1A_1$, (np_x)	
			1.05	4–9	$^1B_1$, (ns)	
			0.05	3–8	$^1A_1$, (nd_{xz})	
H$_2$S(D$_2$S)		84,520	1.57	4–10	$^1B_1(p_z)$ or $^1A_1(p_x)$	87, 94
		84,420	2.04	5–14	$^1B_1$, (ns)	
		84,420	2.27	5–8	(nd)?	
			2.13	5–8	(nd)?	
HS(CH$_3$)	methyl mercaptan	76,138	2.05	5 … 13	(ns)	94
		76,197	1.55	4 … 9	(np)	
H$_2$Se(D$_2$Se)		79,703	3.05	6–20	(ns)	94
		79,677	2.15	5–9		
		79,677	2.55	5–8	(np)	
H$_2$Te(D$_2$Te)		73,705	3.95	7–14	(ns)	94
CH$_2$(CHD, CD$_2$)		83,851	0.12	3–6	$^3\Sigma_u^-$, $(nd\pi_g)$	95
CH$_3$(CD$_3$) (δ)		79,392	0.083	3–9	$^2A_1'$, (nda_1')	96
(γ)			0.090	3–9	$^2E''$, (nde'')	
(β)			0.36	3–7	$^2A_1'$, (nsa_1')	
NH$_3$		82,150	1.02	3–5	$^1A_2''$, (nsa_1')	100
C$_2$H$_2$		92,076	0.50	3–10	$^1\Sigma_u^+$, $(n\pi_g)$	105, 106, 107
		91,950	0.95	3–10	$^1\Pi_u$, $(no\sigma_g)$	

$(CH_3)C_2H$	methyl acetylene (propyne)	91,100	0.96	3–12	$^1\Pi_u$, $(n\sigma_g)$	110, 111
		91,240	0.52	3–13	$^1\Sigma_u^+$, $(n\pi_g)$	112
$(C_2H_5)C_2H$	ethyl acetylene (1-Butyne)	83,570	0.98	4–7		107
		83,580	0.57	3–9		
		83,600	0.33	3–10		
HC_4H	diacetylene	82,100	0.92	5, 6 …		115
		82,100	0.55	3, 4 …		
			0.33	3, 4 …		
$(CH_3)C_4(CH_3)$	dimethyl diacetylene	87,060	0.95	3–9		110
		87,020	0.52	3–8		
C_2H_4	ethylene	92,810	0.20	3–7		121, 120
		84,750	1.09	3–7	$B_{3u}(ns)$	
		84,750	0.60	3–7		
		84,750	0.30	3–7	(nd)	
$(CH_3)HC_2H_2$	propylene	78,586	(0.94)	(3)		120
		(+527)	0.15	4–9		122, 121
C_4H_6	1,3 butadiene	73,115	0.10	4–8	(nd)	123
		73,066	0.50	4–8	(np)	
C_5H_8	methyl butadiene 1,3(2)	71,337	0.10			123
			0.50			
C_6H_{10}	β,γ dimethyl butadiene 1,3	71,245	0.10		as for C_5H_8	
			0.50			
C_4H_5Cl	chlorobuta 1,3-diene	71,203	0.10		as for C_5H_8	
			0.50			
C_3H_4	allene	82,210	1.06	3–8	(ns)	124
		82,190	0.70	3–7	(np)	
		82,190	0.55	3–7	(np)	
		82,200	0.40	3–7	$(nd)?$	
		82,210	0.30	3–6	$(nd)?$	

Chemical Symbol	Name of Species	Ionization Potential (cm^{-1})	δ	Range of n	Upper States excited orbital	References
		(other incomplete series)				
$Cl_2C_2H_2$	1,1-dichloroethylene	76,293	0.93	3–9	(ns)	125
$ClHC_2HCl$	trans dichloro-ethylene	83,630	0.60	3–6	orbital localized on Cl	126
		80,285	0.72	4–9	(ns)?	
$HClC_2HCl$	cis-dichloroethylene	77,937	0.95	3–7	$(ns) \leftarrow \pi_{CC}$	126
		77,850	0.07	3–7	(nd)	
		78,130	1.01	4–7	(ns)	
		77,703	0.045	3–6	(nd)	
Cl_2C_2HCl	trichloroethylene	70,890	0.90	4–9	(ns)	126
$ClHC_2H_2$	vinyl chloride	80,700	0.85	3–9	(ns)	126
		80,645	0.05	3–11	(nd)	
$C_6H_6(C_6D_6)$	benzene	74,590	0.45	4–8		127
		74,495	0.03	4–9		
		74,587	0.46	3–10	$np\pi \leftarrow \pi$	129
			1.16	4–12	$np\sigma \leftarrow \pi$	
			0.11	3–9		
			0.04	3–9	$(nd\delta) \leftarrow \pi$	
$C_6H_5CH_3$	toluene	71,180	0.50	4–7	$(np\pi)$	130
		71,130	0.95	4–9	$(np\sigma)$	
$C_6H_5(C_2H_5)$	ethyl benzene	70,750	1.10	3–5	$(np\sigma)$	
$C_6H_4(CH_3)_2$	O-xylene	69,200	1.08	4–7	$(np\sigma)$	131
C_6H_5F		74,304	0.50	3–8	$(np\pi) \leftarrow \pi$	
		74,203	1.05	3–11	$(np\sigma)$	

Formula	Name				Assignment	Ref.
$C_6H_3F_3$		78,250	0.50	3–7	$(np\pi) \leftarrow \pi$	133
		78,120	1.05	3–11	$(np\sigma)$	
C_5H_5N	pyridine	93,240			(no details, but series similar to benzene)	133
		93,240				
		83,000				
		74,740				
$C_4H_4N_2$	1,4 pyrazine	74,942				134, 133
C_5H_6	1,3 cyclopentadiene	69,550	0.72	5–8		115, 135
C_4H_4O	furan	73,080	0.10	4–7		115
		73,020	0.50	5–7		
		71,678	0.06	3–11	(ns) or $(nd) \leftarrow \pi$	136
		72,747	0.06	3–8	to vibrational state of ion	
		71,688	0.82	4–8		
		80,229	0.28	4–10	from π or N on O-atom	
C_4H_4S	thiophene	72,170	0.10	5–8		115
CH_3I	methyl iodide	76,930	0.25	7–14	A_1 or E 5 $p\pi$(e)	139a
		81,990	0.20	7–14	as above to $^2\pi_{1/2}$ of ion	
CH_3Br		85,020	0.10	4–11		
		87,560	0.10	4–10		
CH_3Cl		90,500	0.50	3–10		
		91,180	0.50	3–11		
C_2H_5I	ethyl iodide	75,380	0.20	5–10		139b
		80,080	0.20	5–10		
CH_3F		104,200	1.05	3,4	(ns), $E \leftarrow \pi$(e)	141
H_2CCl_2		90,500	0.50	4–10		143
H_2CO	formaldehyde	87,809	1.04	3–8	(ns), $B_2 \leftarrow N$	118, 94, 144
		87,710	0.70	3–9	(np_z, p_y), B_2, A_1	
		87,830	0.40	3–8	$(nd) \leftarrow N$	

Chemical Symbol	Name of Species	Ionization Potential (cm⁻¹)	δ	Range of n	Upper States excited orbital	References
CH_3HCO	acetaldehyde	82,504	0.90	3–10	as for H_2CO	145
		82,505	0.70	3–18		
		82,475	0.20	3–10		
$(CH_3)_2CO$	acetone	82,767	0.03	3–12	(nd)	146
		78,280				147
C_2H_4CO	acrolein	81,460	0.95	4–8	$(ns) \leftarrow N$	148
		81,516	0.68	3–12	(np)	
		81,500	0.15	3–10	(nd)	
H_2CCO	ketene	77,491	1.07	3–8	$(ns) \leftarrow N$	121, 149
C_3O_2	carbon suboxide	85,500	1.00	3–10	$(ns) \leftarrow (\pi_u)$	150
		85,500	0.76	3–7	$(n\pi_g) \leftarrow (\pi_u)$	
H_2CNN	diazomethane	72,585	0.10	3–10	(nd) or (ns)	151
C_4H_6O	cyclobutanone	75,444	1.16	4–8	(ns)	152
$HCOOH$	formic acid	91,370	0.60	3–6	$(np) \leftarrow N$	153
$(CH_3)_2O$	dimethyl ether	80,330	0.02	3–10	(nd)	154
		$+r\nu_7$; $r = 0, 1, 2, 3$ $\nu_7 = 450$ cm⁻¹				
$(CH_2)_3O$	trimethylene oxide	77,980	0.05	3–10	$(nd) \leftarrow N$	155
		78,130	0.05	3–7	vibrational series	
		79,330		3–6	with $\nu = 150$–200 cm⁻¹	
C_5H_8O	tetrahydrofuran	75,960	0.10	3–8	$(nd) \leftarrow N$	155
		76,140	0.10	3–6	vibrational series $\nu = 180$	
		77,580	0.20	3–6	$(nd) \leftarrow N$	
		77,720	0.20	3–5	vibrational series $\nu = 140$	

Formula	Name	ν				Ref.
$C_5H_{10}O$	tetrahydropyran	74,630	0.10	3–8	$(nd) \leftarrow N$	155
C_2H_4O	ethylene oxide	87,236	0.55	3–10	(as acetylene)	156
		87,175	0.95	4–9	$(nd) \leftarrow N$	157, 158
		85,200	0.04	3–8	vibrational series	
		86,350	0.04	3–8	$\nu' = 1150$	
$C_4H_8O_2$	1,4 dioxane	76,750	0.13	3–5	series doubtful	159
CO_2		111,060	0.65	3–15	converges to $^2\Pi_{g3/2}$ of CO_2^+	164, 165, 166
		111,240	0.65	3–15	to $^2\Pi_{g1/2}$ of CO_2^+	
		111,250	0.57	3–11		
		111,250	0.97	4–11		
		$+\nu' = 1260$	0.70	5–11		
		139,634	0.04	6–9	to $^2\Pi_{u3/2}$ of CO_2^+	
		$+\nu' = 1130$	0.04	5–9		
		$+2\nu'$	0.04	5–9		
		$+3\nu'$	0.04	5–9		
		$+4\nu'$	0.04	4–7		
		$+5\nu'$	0.04	4–7		
		$+6\nu', 7\nu'$	0.04	4–5, 4–5		
		139,726	0.093	3–10	to $^2\Pi_{u1/2}$ of CO_2^+	
		$+\nu' = 1130$	0.093	3–10		
		$+2\nu'$	0.093	3–10		
		$+3\nu'$	0.093	3–10	strongest series	
		$+4\nu'$	0.093	3–10		
		$+(5,6,7\nu')$		3–8, 4–8, 4–6		
		145,800	0.90	4–19	to $^1\Sigma_u^+$ of CO_2^+	166, 89
		145,800	0.30	3–18		
		$+\nu' = 1250$	0.10	8–11		
		156,350	0.71	4–9	to $^1\Sigma_g^+$ of CO_2	164, 165

Chemical Symbol	Name of Species	Ionization Potential (cm⁻¹)	δ	Range of n	Upper States excited orbital	References
CO_2		156,400	0.05	3–8		
		156,410	0.56	4–8		170, 171
CS_2		81,299	0.44	4–19	to $^2\Pi_{g3/2}$ of CS_2^+	
		81,735	0.46	4–19	to $^2\Pi_{g1/2}$ of CS_2^+	
		116,760	0.94	3–14	to $^2\Sigma_u^+$ of CS_2^+	
		116,760	0.58	3–8	to $^2\Sigma_u^+$ of CS_2^+	
		130,540	0.81	4–13	to $^2\Sigma_g^+$ of CS_2^+	
		130,600	0.10	3–9	to $^2\Sigma_g^+$ (apparent emission)	
		130,610	0.48	3–7	to $^2\Sigma_g^+$ (apparent emission)	
		157,390	0.95	4–11	to 4th level of CS_2^+	
COS		90,610	0.38	5–11	probably to $^2\Pi$ of COS^+	171, 167
		90,740	0.44	6–11		
		129,400	0.68	3–13	to $^2\Sigma$ of COS^+	
		129,400	0.49	3–8		
		129,390	0.12	3–9		
		129,330	0.96	4–8		
		144,680	0.90	5–10	probably to higher $^2\Sigma$	
NNO	nitrous oxide	102,567	0.92	5–8	(reanalysis of 172)	171
		104,000	0.60	3–13	to $^2\Pi$ upper level	
		103,300	0.68	3–9		
		132,210	1.00	3–12	to $^2\Sigma$	

Molecule					Ref.
NNO	132,250	0.22	3-8		
	+ν' = 1288.5	0.11	8-11		
NO₂ (nitrogen dioxide)	162,130	0.31	3-11		174, 176
	162,200	0.06	3-11		175
	162,200	0.68	3-6		
	162,160	0.58	3-6		
	99,500	0.75	3,4(5,6)	from (3b₂) or (1a₁)	
	93,695	0.79	(4),5-9		
	(78,895)	no series observed from (4a₁)			
	152,100	0.15	2-8	from (2a₁)	176
	152,290	0.67	3-7		
	152,160	0.03	2-4		
	+ν' = 230	0.03	2-4		
	112,770	no series			
SO₂	99,500	0.02	3-9	(or δ = 1.02, n 4-10)	170, 177
	+ν' = 900	0.02	3-9		
	+2ν'	0.02	3-9		
XeF₂	92,000	0.98	3,4	spin orbit coupling	180
	98,000	0.98	3,4	gives two limits	

References

1. A. Fowler, "Report on Series in Line Spectra." Fleetway Press, London 1922.
2. F. Paschen and R. Götze, "Seriengesetz der Linienspectren." Springer, Berlin, 1922.
3. L. Pauling and L. Goudsmit, "The Structure of Line Spectra." McGraw-Hill, New York, 1930.
4. C. E. Moore, Atomic energy levels. *Nat. Bur. Stand. (U.S.) Circ.* **467**, Vol. I (^{1}H-^{23}V) (1949); Vol. II (^{24}Cr-^{41}Nb) (1952); Vol. III (^{42}Mo-^{89}Ac) (1958).
5. N. Bohr, *Phil. Mag.* **26**, 1 (1913).
6a. R. S. Mulliken, *J. Amer. Chem. Soc.* **86**, 3183 (1964).
6b. R. S. Mulliken, *J. Amer. Chem. Soc.* **88**, 1849 (1966).
6c. R. S. Mulliken, *J. Amer. Chem. Soc.* **91**, 4615 (1969).
7. G. Herzberg, "Molecular Structure and Molecular Spectra," Vol. I, "Spectra of Diatomic Molecules." Van Nostrand, Princeton, New Jersey, 1950.
8a. H. A. Bethe and E. E. Salpeter, "Quantum Mechanics of One- and Two-Electron Atoms." Academic Press, New York, 1957.
8b. H. A. Bethe and E. E. Salpeter, "Quantum Mechanics of One- and Two-Electron Atoms," p. 123. Academic Press, New York, 1957.
9. J. K. L. MacDonald, *Phys. Rev.* **43**, 830 (1933).
10. L. Pauling, and E. B. Wilson, Jr., "Introduction to Quantum Mechanics," p. 186. McGraw-Hill, New York, 1936.

11. C. Eckart, *Phys. Rev.* **36,** 878 (1930).
12. H. Shull and P.–O. Löwdin, *Phys. Rev.* **110,** 1466 (1958).
13. O. Sinanoğlu, *Phys. Rev.* **122,** 491 (1961).
14. V. Fock, *Z. Phys.* **61,** 126 (1930).
15. D. R. Hartree, "The Calculation of Atomic Structures." Wiley, New York, 1957.
16. J. C. Slater, "Quantum Theory of Atomic Structure," Vol. 1, Chapter 9. McGraw-Hill, New York, 1960.
17. C. C. J. Roothaan, *Rev. Mod. Phys.* **23,** 69 (1951).
18. C. C. J. Roothaan, and P. S. Bagus, Methods in computational physics. *In* "Quantum Mechanics" (B. Alder, ed.), Vol. 2, p. 47. Academic Press, New York, 1963.
19. H. A. Bethe, and E. E. Salpeter, "Quantum Mechanics of One- and Two-Electron Atoms," Table 4, p. 136. Academic Press, New York, 1957.
20. L. P. Smith, *Phys. Rev.* **42,** 176 (1932).
21. E. A. Hylleraas, *Z. Phys.* **48,** 469 (1928); **54,** 347 (1929).
22. G. Herzberg, and R. Zbinden, see H. A. Bethe and E. E. Salpeter, "Quantum Mechanics of One- and Two-Electron Atoms," p. 153. Academic Press, New York, 1957.
23. G. R. Taylor and R. G. Parr, *Proc. Nat. Acad. Sci. U.S.* **38,** 154 (1952).
24. E. A. Hylleraas and B. Underheim, *Z. Phys.* **65,** 759 (1930).
25. A. S. Coolidge and H. M. James, *Phys. Rev.* **49,** 676 (1936).
26. C. A. Coulson and C. S. Sharma, *Proc. Roy. Soc. Ser. A* **272,** 1 (1963).
27. M. Cohen and P. S. Kelly, *Can. J. Phys.* **43,** 1867 (1965).
28. J. Linderberg, *Phys. Rev.* **121,** 816 (1961).
29. Y. Öhrn and R. McWeeny, *Ark. Fys.* **31,** 461 (1966).
30. P. G. Lykos and R. G. Parr, *J. Chem. Phys* **24,** 1166 (1956).
31. R. McWeeny, *Proc. Roy. Soc. Ser. A* **223,** 306 (1954).
32. A. U. Hazi and S. A. Rice, *J. Chem. Phys.* **45,** 3004 (1966); references here to many earlier articles.
33. C. A. Coulson and J. G. Stamper, *Mol. Phys.* **6,** 609 (1963).
34. C. C. J. Roothaan, L. M. Sachs, and A. W. Weiss, *Rev. Mod. Phys.* **32,** 186 (1960).
35. D. B. Cook and J. N. Murrell, *Mol. Phys.* **9,** 417 (1965).
36. T. F. Lin and A. B. F. Duncan, *J. Chem. Phys.* **51,** 360 (1969).
37. D. F. Tuan and O. Sinanoğlu, *J. Chem. Phys.* **41,** 2677 (1964).
38. M. Kotani, A formal theory of Rydberg series of molecules. *In* "Molecular Orbitals in Chemistry, Physics and Biology. A Tribute to R. S. Mulliken," p. 539. Academic Press, New York, 1964. Edited by P.-O. Löwdin and B. Pullman.
39. D. R. Bates, K. Ledsham, and A. L. Stewart, *Phil. Trans. Roy. Soc. London Ser. A* **246,** 215 (1954).
40. G. Herzberg, "Molecular Structure and Molecular Spectra," Vol. I, "Spectra of Diatomic Molecules," pp. 328, 329. Van Nostrand, Princeton, New Jersey, 1950.
41. S. Takezawa, *J. Chem. Phys.* **52,** 2575, 5793 (1970).

42. F. A. Matsen and J. C. Browne, Rydberg orbitals and energies for H_2. *In* "Molecular Orbitals in Chemistry, Physics and Biology. A Tribute to R. S. Mulliken," p. 151. Academic Press, New York, 1964. Edited by P.-O Löwdin and B. Pullman.
43. A. U. Hazi and S. A. Rice, *J. Chem. Phys.* **47**, 1125 (1967).
44. G. Herzberg, "Molecular Structure and Molecular Spectra," Vol. I, "Spectra of Diatomic Molecules," pp. 535, 536. Van Nostrand, Princeton, New Jersey, 1950. See also W. Jevons, "Report on Band-Spectra of Diatomic Molecules." Phys. Soc., Cambridge Univ. Press, London and New York, 1932.
45. R. S. Mulliken, *Rev. Mod. Phys.* **2**, 60 (1930).
46. M. L. Ginter, *J. Chem. Phys.* **44**, 950 (1966).
47. M. L. Ginter, *J. Chem. Phys.* **42**, 561 (1965); *J. Mol. Spectrosc.* **17**, 224 (1965); **18**, 321 (1965).
48. R. S. Mulliken, *Phys. Rev. A* **136**, 962 (1964).
49a. R. E. Worley and F. A. Jenkins, *Phys. Rev.* **54**, 305 (1938).
49b. R. E. Worley, *Phys. Rev.* **64**, 207 (1943).
50. R. E. Worley, *Phys. Rev.* **65**, 249 (1944).
51. P. K. Carroll, *J. Quant. Spectrosc. Radiat. Transfer* **2**, 417 (1962).
52. M. Ogawa and Y. Tanaka, *Can. J. Phys.* **40**, 1593 (1962).
53. P. G. Wilkinson and N. B. Houk, *J. Chem. Phys.* **24**, 528 (1956).
54. K. Dressler, *Can. J. Phys.* **47**, 547 (1969).
55. A. von Lagerqvist and E. Miescher, *Helv. Phys. Acta* **31**, 221 (1958).
56. R. E. Huffman, J. C. Larrabee and Y. Tanaka, *J. Chem. Phys.* **45**, 3205 (1966).
57. R. E. Worley, *Phys. Rev.* **89**, 863 (1953).
58. J. J. Hopfield, *Phys. Rev.* **35**, 1133; **36**, 789 (1930).
59. Y. Tanaka, *Sci. Pap. Inst. Phys. Chem. Res. Tokyo* **39**, 456 (1942).
60. T. Takamine, Y. Tanaka and M. Iwata, *Sci. Pap. Inst. Phys. Chem. Res. Tokyo* **40**, 371 (1943).
61. Y. Tanaka, *Sci. Pap. Inst. Phys. Chem. Res. Tokyo* **39**, 465 (1942).
62. K. Watanabe, *J. Chem. Phys.* **22**, 1564 (1954); **26**, 542 (1957).
63. D. W. Turner and D. P. May, *J. Chem. Phys.* **45**, 471 (1966).
64. Y. Tanaka, *J. Chem. Phys.* **21**, 562 (1953).
65. E. Miescher and P. Baer, *Nature (London)* **169**, 581 (1952).
66. P. G. Wilkinson, *J. Mol. Spectrosc.* **6**, 1 (1961).
67. K. Yoshino and Y. Tanaka, *J. Chem. Phys.* **48**, 4859 (1968).
68. E. Lindholm, *Ark. Fys.* **40**, 111 (1970).
69. W. C. Price and G. Collins, *Phys. Rev.* **48**, 714 (1935).
70. Y. Tanaka and T. Takamine, *Sci. Pap. Inst. Phys. Chem. Res. Tokyo* **39**, 437 (1942).
71. K. Codling and R. P. Madden, *J. Chem. Phys.* **42**, 3935 (1965).
72. P. Venkateswarlu, *Can. J. Phys.* **47**, 2525 (1969).
73. P. Venkateswarlu, *Can. J. Phys.* **48**, 1055 (1970).
74. W. C. Price, *Proc. Roy. Soc. Ser. A* **167**, 216 (1938).
75. R. S. Mulliken, *Phys. Rev.* **61**, 277 (1942).

76. G. Herzberg, "Molecular Structure and Molecular Spectra," Vol. I, "Spectra of Diatomic Molecules," p. 342. Van Nostrand, Princeton, New Jersey, 1950.

77. R. B. Caton and A. E. Douglas, *Can. J. Phys.* **48**, 432 (1970).

78. G. Herzberg, "Molecular Structure and Molecular Spectra," Vol. I, "Spectra of Diatomic Molecules," p. 228. Van Nostrand, Princeton, New Jersey, 1950.

79. S. H. Bauer, G. Herzberg and J. W. C. Johns, *J. Mol. Spectrosc.* **13**, 256 (1964).

80. G. Herzberg and J. W. C. Johns, *Astrophys. J.* **158**, 399 (1969).

81. H. Lefebvre-Brion, C. M. Moser, and R. K. Nesbet, *J. Mol. Spectrosc.* **15**, 211 (1965); BF, CO.

82. H. Lefebvre-Brion and C. M. Moser, *J. Mol. Spectrosc.* **13**, 418 (1964); CO.

83. H. Lefebvre-Brion and C. M. Moser, *J. Chem. Phys.* **43**, 1394 (1965); N_2.

84. A. U. Hazi and S. A. Rice, *J. Chem. Phys.* **48**, 495 (1968).

85. Y. Kato, E. F. Hayes and A. B. F. Duncan, *J. Chem. Phys.* **41**, 986 (1964).

86. A. B. F. Duncan and A. Damiani, *J. Chem. Phys.* **45**, 1245 (1966); **42**, 2453 (1965).

87. W. C. Price, *J. Chem. Phys.* **4**, 147 (1936).

88. J. W. C. Johns, *Can. J. Phys.* **39**, 1738 (1961).

89. H. J. Henning, *Ann. Phys. New York* **13**, 599 (1932).

90. D. C. Frost and C. A. McDowell, *Can. J. Chem.* **36**, 39 (1958).

91. T. F. Lin and A. B. F. Duncan, *J. Chem. Phys.* **48**, 866 (1968).

92. S. R. LaPaglia, *J. Chem. Phys.* **41**, 1427 (1964).

93. Y. Harada and J. N. Murrell, *Mol. Phys.* **14**, 153 (1968).

94. W. C. Price, J. P. Teegan and A. D. Walsh, *Proc. Roy. Soc. Ser. A* **201**, 600 (1950).

95. G. Herzberg, *Can. J. Phys.* **39**, 1511 (1961); G. Herzberg and J. Shoosmith, *Nature (London)* **183**, 1801 (1959).

96. G. Herzberg, *Proc. Roy. Soc. Ser. A* **262**, 291 (1961).

97. G. Herzberg and J. Shoosmith, *Can. J. Phys.* **34**, 523 (1956).

98. A. E. Douglas and J. M. Hollas, *Can. J. Phys.* **39**, 479 (1961).

99. A. E. Douglas, *Discuss. Faraday Soc.* **35**, 158 (1963).

100. A. D. Walsh and P. A. Warsop, *Trans. Faraday Soc.* **57**, 345 (1961).

101. A. B. F. Duncan, *Phys. Rev.* **47**, 822 (1935); **50**, 700 (1936).

102. K. Watanabe and J. R. Mottl, *J. Chem. Phys.* **26**, 1773 (1957).

103. A. D. Walsh and P. A. Warsop, *Advan. Mol. Spectrosc., Proc. Int. Meet. 4th, 1959* (A. Mangini, ed.), **2**, Permagon, New York, 1962.

104. G. L. Humphries, A. D. Walsh and P. A. Warsop, *Discuss. Faraday Soc.* **35**, 148 (1963).

105. W. C. Price, *Phys. Rev.* **47**, 444 (1935).

106. P. G. Wilkinson, *J. Mol. Spectrosc.* **2**, 387 (1958).

107. T. Nakayama and K. Watanabe, *J. Chem. Phys.* **40**, 558 (1964).

108. G. Herzberg, *Discuss. Faraday Soc.* **35**, 7 (1963).

109. E. Greene, Jr., J. Barnard, and A. B. F. Duncan, *J. Chem. Physic* **54** 000 (1971).

110. W. C. Price and A. D. Walsh, *Trans. Faraday Soc.* **41,** 381 (1945).
111. Quoted and refered to in Ref. 112 by T. Namioka and K. Watanabe.
112. T. Namioka and K. Watanabe, *J. Chem. Phys.* **24,** 915 (1956).
113. J. L. Franklin and F. H. Field, *J. Amer. Chem. Soc.* **76,** 1994 (1954).
114. F. H. Coates and R. C. Andersen, *J. Amer. Chem. Soc.* **77,** 895 (1955).
115. W. C. Price and A. D. Walsh, *Proc. Roy. Soc. Ser. A* **179,** 201 (1941).
116. J. H. Callomon, *Can. J. Phys.* **34,** 1046 (1956).
117. C. A. McDowell and J. W. Warren, *Trans. Faraday Soc.* **48,** 1084 (1952).
118. G. Herzberg, "Molecular Spectra and Molecular Structure," Vol. III, "Electronic Spectra and Electronic Structure of Polyatomic Molecules." Van Nostrand, Princeton, New Jersey, 1966.
119. P. G. Wilkinson and R. S. Mulliken, *J. Chem. Phys.* **23,** 1895 (1953).
120. P. G. Wilkinson, *Can. J. Phys.* **34,** 643 (1956).
121. W. C. Price and W. T. Tutte, *Proc. Roy. Soc. Ser. A* **174,** 207 (1940).
122. J. A. R. Samson, F. F. Marmo and K. Watanabe, *J. Chem. Phys.* **36,** 783 (1962).
123. W. C. Price and A. D. Walsh, *Proc. Roy. Soc. Ser. A* **174,** 220 (1940).
124. L. H. Sutcliffe and A. D. Walsh, *J. Chem. Soc.* p. 899 (1952).
125. J. P. Teegan and A. D. Walsh, *Trans. Faraday Soc.* **47,** 1 (1951).
126. A. D. Walsh, *Trans. Faraday Soc.* **41,** 35 (1945); see also H. E. Mahncke and W. A. Noyes, Jr., *J. Chem. Phys.* **3,** 536 (1935).
127. W. C. Price and R. W. Wood, *J. Chem. Phys.* **3,** 439 (1935).
128. M. A. El-Sayed, M. Kasha and Y. Tanaka, *J. Chem. Phys.* **34,** 334 (1961).
129. P. G. Wilkinson, *Can. J. Phys.* **34,** 596 (1956).
130. W. C. Price and A. D. Walsh, *Proc. Roy. Soc. Ser. A* **191,** 22 (1947).
131. V. J. Hammond, J. P. Teegan and A. D. Walsh, *Discuss. Faraday Soc.* **9,** 53 (1950).
132. K. Watanabe, *J. Chem. Phys.* **26,** 542 (1957).
133. M. A. El-Sayed, *J. Chem. Phys.* **36,** 552 (1962).
134. J. E. Parkin and K. K. Innes, *J. Mol. Spectrosc.* **15,** 407 (1965).
135. L. W. Pickett, *J. Chem. Phys.* **8,** 293 (1940); L. W. Pickett, N. J. Hoeflich and T. C. Liu, *J. Amer. Chem. Soc.* **73,** 4865 (1951).
136. K. Watanabe and T. Nakayama, *J. Chem. Phys.* **29,** 48 (1958).
137. A. B. F. Duncan and J. P. Howe, *J. Chem. Phys.* **2,** 851 (1934); G. Moe and A. B. F. Duncan, *J. Amer. Chem. Soc.* **74,** 3140 (1952); N. Wainfan, W. C. Walker, and G. L. Weissler, *Phys. Rev.* **99,** 542 (1955); H. Sun and G. L. Weissler, *J. Phys. Chem.* **23,** 1160 (1955).
138. P. H. Metzger and G. R. Cook, *J. Chem. Phys.* **41,** 642 (1964).
139a. W. C. Price, *J. Chem. Phys.* **4,** 539 (1936).
139b. W. C. Price, *J. Chem. Phys.* **4,** 547 (1936).
140. R. S. Mulliken, *Phys. Rev.* **47,** 413 (1935).
141. S. Stokes, and A. B. F. Duncan, *J. Amer. Chem. Soc.* **80,** 6177 (1958).
142. C. A. McDowell and B. C. Cox, *J. Chem. Phys.* **22,** 946 (1954).
143. C. R. Zobel and A. B. F. Duncan, *J. Amer. Chem. Soc.* **77,** 2611 (1955).
144. K. Allison and A. D. Walsh, The structure and reactivity of electronically

excited species. *Chem. Inst. Can. Sympo., September 1957.* Univ. of Ottawa, Canada, 1957.

145. A. D. Walsh, *Proc. Roy. Soc. Ser. A* **185,** 176 (1946).
146. A. B. F. Duncan, *J. Chem. Phys.* **3,** 131 (1935).
147. K. Watanabe, *J. Chem. Phys.* **22,** 1564 (1954).
148. A. D. Walsh, *Trans. Faraday Soc.* **41,** 498 (1945).
149. W. C. Price, J. P. Teegan, and A. D. Walsh, *J. Chem. Soc.* p. 920 (1951).
150. J. L. Roebber, J. C. Larrabee, and R. E. Huffman, *J. Chem. Phys.* **46,** 4594 (1967).
151. A. J. Merer, *Can. J. Phys.* **42,** 1242 (1964).
152. R. F. Whitlock and A. B. F. Duncan, *J. Chem. Phys.* **54** (1971)
153. W. C. Price and W. M. Evans, *Proc. Roy. Soc. Ser. A* **162,** 110 (1937).
154. G. J. Hernandez, *J. Chem. Phys.* **38,** 1644 (1963).
155. G. J. Hernandez, *J. Chem. Phys.* **38,** 2233 (1963).
156. T. K. Liu and A. B. F. Duncan, *J. Chem. Phys.* **17,** 241 (1949).
157. A. Lowrey and K. Watanabe, *J. Chem. Phys.* **28,** 208 (1958).
158. H. Basch, M. B. Robin, N. A. Kuebler, C. Baker, and D. W. Turner, *J. Chem. Phys.* **51,** 52 (1969).
159. G. J. Hernandez and A. B. F. Duncan, *J. Chem. Phys.* **36,** 1504 (1962).
160. K. Watanabe, T. Nakayama, and J. Mottl, Final report on ionization potential of molecules by a photoionization method. Dept. of Phys., Univ. of Hawaii, December 1959.
161. T. Lyman, *Astrophys. J.* **27,** 87 (1908).
162. S. W. Leifson, *Astrophys. J.* **63,** 73 (1926).
163. G. Rathenau, *Z. Phys.* **87,** 32 (1934).
164. Y. Tanaka, A. S. Jursa, and F. J. LeBlanc, *J. Chem. Phys.* **28,** 350 (1958).
165. Y. Tanaka, A. S. Jursa, and F. J. LeBlanc, *J. Chem. Phys.* **32,** 1199 (1960).
166. Y. Tanaka and M. Ogawa, *Can. J. Phys.* **40,** 879 (1962).
167. W. C. Price and D. M. Simpson, *Proc. Roy. Soc. Ser. A* **169,** 501 (1939).
168. E. C. Y. Inn, K. Watanabe, and M. Zelicoff, *J. Chem. Phys.* **21,** 1648 (1953).
169. P. G. Wilkinson and H. L. Johnston, *J. Chem. Phys.* **18,** 190 (1950).
170. W. C. Price and D. M. Simpson, *Proc. Roy. Soc. Ser. A* **165,** 272 (1938).
171. Y. Tanaka, A. S. Jursa, and F. J. LeBlanc, *J. Chem. Phys.* **32,** 1205 (1960).
172. A. B. F. Duncan, *J. Chem. Phys.* **4,** 638 (1936).
173. M. Zelikoff, K. Watanabe, and E. C. Y. Inn, *J. Chem. Phys.* **21,** 1643 (1953).
174. W. C. Price and D. M. Simpson, *Trans. Faraday Soc.* **37,** 106 (1941).
175. T. Nakayama, M. Y. Kitamura, and K. Watanabe, *J. Chem. Phys.* **30,** 1180 (1959).
176. Y. Tanaka and A. S. Jursa, *J. Chem. Phys.* **36,** 2493 (1962).
177. D. Golomb, K. Watanabe, and F. F. Marmo, *J. Chem. Phys.* **36,** 958 (1962).
178. Y. Tanaka, E. C. Y. Inn, and K. Watanabe, *J. Chem. Phys.* **21,** 1651 (1953).
179. M. Ogawa and G. R. Cook, *J. Chem. Phys.* **28,** 173 (1958).
180. E. G. Wilson, J. Jortner, and S. A. Rice, *J. Amer. Chem. Soc.* **85,** 813 (1963).
181. J. Jortner, E. G. Wilson, and S. A. Rice, *J. Amer. Chem. Soc.* **85,** 815 (1963).

REFERENCE AND AUTHOR INDEX

Numbers in parentheses indicate the superscript reference numbers cited in text; numerals following indicate pages on which the reference is cited; italic numbers refer to the pages on which the *complete* references are found.

A

Allison, K. (144), 81, 99, *109*
Andersen, R. C. (114), 71, *109*

B

Baer, P. (65), 50, *107*
Bagus, P. S. (18), 18, *106*
Baker, C. (158), 87, 101, *110*
Barnard, J. (109), 69, *108*
Basch, H. (158), 87, 101, *110*
Bates, D. R. (39), 38, 39, *106*
Bauer, S. H. (79), 58, *108*
Bethe, H. A. (8a, 8b, 19), 10, 12, 19, 21, 22, *105*, *106*
Bohr, N. (5), 3, *105*
Browne, J. C. (42), 42, *107*

C

Callomon, J. H. (116), 72, *109*
Carroll, P. K. (51), 46, *107*
Caton, R. B. (77), 57, *108*
Coates, F. H. (114), 71, *109*
Codling, K. (71), 53, *107*
Cohen, M. (27), 27, *106*
Collins, G. (69), 52, *107*
Cook, D. B. (35), 30, *106*
Cook, G. R. (138, 179), 80, 94, *109*, *110*
Coolidge, A. S. (25), 26, *106*
Coulson, C. A. (26, 33), 26, 29, 67, *106*
Cox, B. C. (142), 81, *109*

D

Damiani, A. (86), 60, *108*
Douglas, A. E. (77, 98, 99), 57, 67, 68, *108*
Dressler, K. (54), 47, *107*
Duncan, A. B. F. (36, 85, 86, 91, 101, 109, 137, 141, 143, 146, 152, 156, 159, 172), 31, 60, 61, 65, 68, 69, 80, 81, 83, 85, 86, 87, 92, 99, 100, 101, *106*, *108*, *109*, *110*

E

Eckart, C. (11), 15, *106*
El-Sayed, M. A. (128, 133), 76, 78, 79, 99, *109*
Evans, W. M. (153), 85, 100, *110*

F

Field, F. H. (113), 71, *109*
Fock, V. (14), 17, 23, *106*
Fowler, A. (1), 1, *105*
Franklin, J. L. (113), 71, *109*
Frost, D. C. (90), 65, *108*

G

Ginter, M. L. (46, 47), 44, *107*
Götze, R. (2), 1, *105*
Golomb, D. (177), 94, 103, *110*
Goudsmit, L. (3), 1, *105*
Greene, E., Jr. (109), 69, *108*

SUBJECT INDEX

A

Acetaldehyde (CH_3HCO), 82, 83, 100
Acetone [$(CH_3)_2CO$], 83, 100
Acetylene (C_2H_2), 69, 74, 96
  comparison with N_2, 70
  configuration, 69
  normal, 69
  theoretical calculations, 69
Acrolein (C_2H_4CO), 83, 100
Aldehydes, 81–85
Alkali atoms, neutral, 6
Allene (C_3H_4), 74, 75, 84, 97
  symmetry group, 74
Ammonia (NH_3), 67, 96
  excited states of, 68
Angular correlation, 23
AsH_3, 68
  deuterated, 68
Atomic orbitals, 4
Atomic spectra, Rydberg series in, 9–33
Atomic units
  of distance, 3
  of energy (Hartree), 3

B

B_2, 45
BF, 45, 56–58
BH, 56, 58
BN, 45
Be, 31, 61
Be^{3+}, energy formulas, 4
Be_2, 45
BeH, 45
BeO, 45

Benzene (C_6H_6), 76–80, 98
  derivatives of, 76–80
  deuterated, 98
Benzotrifluoride, 78
Bohr theory of hydrogen atom, 3
Born–Oppenheimer approximation, 36
Born–Oppenheimer separability, 7
Bromine (Br_2), 54
1,3-Butadiene (C_4H_6), 74, 97
1-Butyne, see Ethyl acetylene

C

C_2, 45
CH, 45, 56, 58
CH_2, 66, 96
  deuterated, 66, 96
  theoretical calculations of lower states, 67
CH_3, 67, 96
  configuration of, 67
  deuterated, 96
  lower excited states of, 67
  normal state of, 67
$C_5H_3F_3$, 99
C_6H_5F, 98
CN, 45
CO, see Carbon monoxide
CO^+
  first excited state of, 50
  lowest state of, 49
  second excited state of, 50
CO_2, see Carbon dioxide
CO_2^+, 89–91
COS, 88, 91, 93, 102
COS^+, 91

Physical Chemistry

A Series of Monographs

Ernest M. Loebl, Editor

Department of Chemistry, Polytechnic Institute of

Brooklyn, Brooklyn, New York

Physical Chemistry

A Series of Monographs